Anibal Maury-Ramirez (Ed.)

Photocalytic Coatings for Air-Purifying, Self-Cleaning and Antimicrobial Properties

MDPI

This book is a reprint of the special issue that appeared in the online open access journal *Coatings* (ISSN 2079-6412) in 2014 and 2015 (available at: http://www.mdpi.com/journal/coatings/special_issues/photocalytic-coatings).

Guest Editor
Anibal Maury-Ramirez
Pontificia Universidad Javeriana Cali
Colombia

Editorial Office
MDPI AG
Klybeckstrasse 64
Basel, Switzerland

Publisher
Shu-Kun Lin

Senior Assistant Editor
Zhiqiao Dong

1. Edition 2015

MDPI • Basel • Beijing • Wuhan

ISBN 978-3-03842-137-5 (PDF)
ISBN 978-3-03842-138-2 (Hbk)

Table of Contents

Chapter 1:
Photocatalytic Removal of Microorganisms

Chapter 2:
Photocatalytic Removal of Dyes and Wettability

Chapter 3:
Photocatalytic Removal of Air Pollutants

Elia Boonen and Anne Beeldens
Recent Photocatalytic Applications for Air Purification in Belgium
Reprinted from: *Coatings* **2014**, *4*(3), 553-573

List of Contributors

Ann-Louise Anderson: School of Engineering and Materials Science, Queen Mary University of London, Mile End Road, London E1 4NS, UK.

Anne Beeldens: Belgian Road Research Center (BRRC), Woluwedal 42, 1200 Brussels, Belgium.

Alexandra Bertron: Université de Toulouse, UPS, INSA, LMDC (Laboratoire Matériaux et Durabilité des Constructions), 135 Avenue de Rangueil, F-31077 Toulouse Cedex 04, France.

Russell Binions: School of Engineering and Materials Science, Queen Mary University of London, Mile End Road, London E1 4NS, UK.

Elia Boonen: Belgian Road Research Center (BRRC), Woluwedal 42, 1200 Brussels, Belgium.

Gwen B. Castillon: Solid State Physics Laboratory, De La Salle University-Manila, 2401 Taft Ave., Manila 1004, Philippines.

Juan D. Cohen: Grupo del Cemento y Materiales de Construción CEMATCO, Universidad Nacional de Colombia, Facultad de Minas, 05001000 Medellín, Colombia.

Kevin Cooke: Teer Coatings Ltd., Miba Coating Group, West Stone House, Berry Hill Industrial Estate, Droitwich WR9 9AS, UK.

Marie Coutand: Université de Toulouse, UPS, INSA, LMDC (Laboratoire Matériaux et Durabilité des Constructions), 135 Avenue de Rangueil, F-31077 Toulouse Cedex 04, France.

Robert F. Cozzens: Chemistry Division, Naval Research Laboratory, 4555 Overlook Avenue SW, Washington, DC 20375, USA.

Elmer Estacio: National Institute of Physics, University of the Philippines, Diliman Quezon City 1101, Philippines.

Akira Fujishima: Photocatalysis International Research Center, Research Institute for Science & Technology, Tokyo University of Science, Noda, Chiba 278-8510, Japan.

Takashi Furuya: Research Center for Development of Far-Infrared Region, University of Fukui, Fukui 910-8507, Japan.

Spencer L. Giles: Chemistry Division, Naval Research Laboratory, 4555 Overlook Avenue SW, Washington, DC 20375, USA.

Caterine Daza Gomez: Science and Technology in Ceramic Materials Group (CYTEMAC), Department of Physics—FACNED, University of Cauca, Popayan 190001, Colombia.

Claire Hill: Cristal Pigment UK Ltd., P.O. Box 26, Grimsby, North East Lincolnshire, DN41 8DP, UK.

Atsushi Iwamae: Research Center for Development of Far-Infrared Region, University of Fukui, Fukui 910-8507, Japan.

Peter Kelly: Faculty of Science and Engineering, Manchester Metropolitan University, Chester Street, Manchester M1 5GD, UK.

Sanjay S. Latthe: Photocatalysis International Research Center, Research Institute for Science & Technology, Tokyo University of Science, Noda, Chiba 278-8510, Japan.

Shanhu Liu: Photocatalysis International Research Center, Research Institute for Science & Technology, Tokyo University of Science, Noda, Chiba 278-8510, Japan.

Robert Louder: Department of Chemistry and Physics, Coastal Carolina University, Conway, SC 29528, USA.

Jeffrey G. Lundin: Chemistry Division, Naval Research Laboratory, 4555 Overlook Avenue SW, Washington, DC 20375, USA.

Santhosh Shimoga Mukunda-Rao: Applied Polymer Materials Laboratory, Department of Chemistry, R. V. College of Engineering, Mysore Road, Bangalore 560059, India.

James C. Moore: Department of Chemistry and Physics, Coastal Carolina University, Conway, SC 29528, USA.

Kazuya Nakata: Photocatalysis International Research Center, Research Institute for Science & Technology, Tokyo University of Science, Noda, Chiba 278-8510, Japan.

Kandasamy Natarajan: Applied Polymer Materials Laboratory, Department of Chemistry, R. V. College of Engineering, Mysore Road, Bangalore 560059, India.

Parnia Navabpour: Teer Coatings Ltd., Miba Coating Group, West Stone House, Berry Hill Industrial Estate, Droitwich WR9 9AS, UK.

Lars Österlund: Department of Engineering Sciences, The Ångström Laboratory, Uppsala University, P.O. Box 534, SE-75121 Uppsala, Sweden.

Soheyla Ostovarpour: Faculty of Science and Engineering, Manchester Metropolitan University, Chester Street, Manchester M1 5GD, UK.

Outi Priha: VTT Technical Research Centre of Finland, P.O. Box 1000, FI-02044 VTT Espoo, Finland.

Mari Raulio: VTT Technical Research Centre of Finland, P.O. Box 1000, FI-02044 VTT Espoo, Finland.

Jorge Enrique Rodriguez-Paez: Science and Technology in Ceramic Materials Group (CYTEMAC), Department of Physics—FACNED, University of Cauca, Popayan 190001, Colombia.

Christine Roques: Université de Toulouse, UPS, LGC (Laboratoire de Génie Chimique), Dép. BioSyM, UFR Pharmacie–35 rue des Maraîchers, 31062 Toulouse Cedex 09, France.

Gil Nonato C. Santos: Solid State Physics Laboratory, De La Salle University-Manila, 2401 Taft Ave., Manila 1004, Philippines.

Germán Alberto Sierra-Gallego: Grupo Investigación en catálisis y nano materiales, Universidad Nacional de Colombia, Facultad de Minas, 05001000 Medellín, Colombia.

Bozhidar Stefanov: Department of Engineering Sciences, The Ångström Laboratory, Uppsala University, P.O. Box 534, SE-75121 Uppsala, Sweden.

Masahiko Tani: Research Center for Development of Far-Infrared Region, University of Fukui, Fukui 910-8507, Japan.

Carin Tattershall: Cristal Pigment UK Ltd., P.O. Box 26, Grimsby, North East Lincolnshire, DN41 8DP, UK.

Chiaki Terashima: Photocatalysis International Research Center, Research Institute for Science & Technology, Tokyo University of Science, Noda, Chiba 278-8510, Japan.

Cody V. Thompson: Department of Chemistry and Physics, Coastal Carolina University, Conway, SC 29528, USA; Research & Development Laboratory, Wellman Engineering Resins, Johnsonville, SC 29555, USA

Eduardo B. Tibayan: Solid State Physics Laboratory, De La Salle University-Manila, 2401 Taft Ave., Manila 1004, Philippines.

Jorge I. Tobón: Grupo del Cemento y Materiales de Construción CEMATCO, Universidad Nacional de Colombia, Facultad de Minas, 05001000 Medellín, Colombia.

Thomas Verdier: Université de Toulouse, UPS, INSA, LMDC (Laboratoire Matériaux et Durabilité des Constructions), 135 Avenue de Rangueil, F-31077 Toulouse Cedex 04, France.

Joanna Verran: Faculty of Science and Engineering, Manchester Metropolitan University, Chester Street, Manchester M1 5GD, UK.

Kathryn Whitehead: Faculty of Science and Engineering, Manchester Metropolitan University, Chester Street, Manchester M1 5GD, UK.

James H. Wynne: Chemistry Division, Naval Research Laboratory, 4555 Overlook Avenue SW, Washington, DC 20375, USA.

Kohji Yamamoto: Research Center for Development of Far-Infrared Region, University of Fukui, Fukui 910-8507, Japan.

About the Guest Editor

Anibal Maury-Ramirez is an Associate Professor from The Civil Engineering Program at Pontificia Universidad Javeriana Cali (Colombia). He received the degree in Civil Engineering with a final research work on "Development of a neural-fuzzy model to estimate the in-let flows to waste treatment plants" (Universidad del Norte - Colombia, 2003). Based on his academic and research performance, he received a scholarship for an international ALFA program on Materials Science in which he investigated "TiO_2 impregnation of concrete and plaster surfaces" (Tampere University of Technology - Finland, 2004). Following the results obtained in the pioneer research work of photocatalytic coatings applied on plasters and concrete, he started a broader research on this topic during his doctorate in the development of "Cementitious materials with air-purifying and self-cleaning properties using titanium dioxide photocatalysis" (Ghent University - Belgium, 2011). Later, he joined a well-known research group on the use of recycled glass as fine aggregate in mortars (The Hong Kong Polytechnic University - Hong Kong, 2014). Thus, he combined TiO_2 photocatalysis experience with the use of recycled glass to enhance photocatalytic activity in cementitious materials. To date, Professor Maury-Ramirez has authored more than 10 journal publications, 20 articles on conference proceedings and several book chapters on the development of photocatalytic building materials.

Preface

This Special Issue of *Coatings* is focused on the study of different photocatalyst-based coatings for developing self-cleaning, air-purifying and antibacterial properties. In this case, a wide variety of photocatalysts (TiO_2, $Si-TiO_2$, $TiO_{2-x}N_y$, $Ag-TiO_2$, $Mo-TiO_2$, ZnO, SnO_2-Ag, Nb_2O_5 and C60 fullerene) were evaluated towards the removal of different molecules. Similarly, substrates such as glass, silica, sapphire, polycarbonate, aluminium, stainless steel, concrete and mortar were included in this issue. This information certainly contributes to a better understanding of the photocatalytic removal of different molecules (e.g., *Escherichia coli*, *Staphylococcus aureus*, resazuring, rhodamine B, methylene blue, Demeton-S, 2-chloroethyl phenyl sulfide (CEPS) and NO_x (NO and NO_2)) and the coating technologies required for such performances. Based on these interesting results, I encourage you to read through this Special Issue and use the valuable information provided therein to help us move forward in the exciting area of photocatalytic coatings for developing air-purifying, self-cleaning, and antimicrobial properties.

Anibal Maury-Ramirez
Guest Editor

Chapter 1:
Photocatalytic Removal of Microorganisms

Antibiofilm Activity of Epoxy/Ag-TiO₂ Polymer Nanocomposite Coatings against *Staphylococcus aureus* and *Escherichia coli*

Santhosh Shimoga Mukunda-Rao and Kandasamy Natarajan

Abstract: Dispersion of functional inorganic nano-fillers like TiO_2 within polymer matrix is known to impart excellent photobactericidal activity to the composite. Epoxy resin systems with Ag^+ ion doped TiO_2 can have combination of excellent biocidal characteristics of silver and the photocatalytic properties of TiO_2. The inorganic antimicrobial incorporation into an epoxy polymeric matrix was achieved by sonicating laboratory-made nano-scale anatase TiO_2 and Ag-TiO_2 into the industrial grade epoxy resin. The resulting epoxy composite had ratios of 0.5–2.0 wt% of nano-filler content. The process of dispersion of Ag-TiO_2 in the epoxy resin resulted in concomitant *in situ* synthesis of silver nanoparticles due to photoreduction of Ag^+ ion. The composite materials were characterized by DSC and SEM. The glass transition temperature (T_g) increased with the incorporation of the nanofillers over the neat polymer. The materials synthesized were coated on glass petri dish. Anti-biofilm property of coated material due to combined release of biocide, and photocatalytic activity under static conditions in petri dish was evaluated against *Staphylococcus aureus* ATCC6538 and *Escherichia coli* K-12 under UV irradiation using a crystal violet binding assay. Prepared composite showed significant inhibition of biofilm development in both the organisms. Our studies indicate that the effective dispersion and optimal release of biocidal agents was responsible for anti-biofilm activity of the surface. The reported thermoset coating materials can be used as bactericidal surfaces either in industrial or healthcare settings to reduce the microbial loads.

Reprinted from *Coatings*. Cite as: M, S.S.; Natarajan, K. Antibiofilm Activity of Epoxy/Ag-TiO₂ Polymer Nanocomposite Coatings against *Staphylococcus Aureus* and *Escherichia Coli*. *Coatings* **2015**, *5*, 95-114.

1. Introduction

Biofilms are defined as communities of microorganisms that are developed on material surfaces. Prevention of microbial biofilm formation over the surface of materials is a technological imperative in health care. Many bacteria capable of forming biofilms on abiotic surfaces are menacing problems in medical and industrial systems. The biofilm forming ability of the opportunistic human pathogens *Staphylococcus aureus* and *Escherichia coli*, is a crucial step for sustenance and growth in above said environments [1]. Biofilms are a major source of biofouling in industrial water systems, and biofilm based industrial slimes also pose major problems for various industrial processes. Biofilm forming microbial cells attached to any surface in a moist environment can survive and proliferate. Pathogenic and resilient biofilms are difficult to eradicate with conventional disinfectants [2]. The interest in inorganic disinfectants such as metal oxide nanoparticles (NPs) is increasing. In the last decade, many studies describing the photocatalytic inactivation of bacteria using doped and undoped TiO_2 coated on different substrates have been reported, including silver doped TiO_2 [3–6]. A majority of these articles is focused on powder materials and thin films of TiO_2 or doped TiO_2. Unfortunately, most bare TiO_2 coated films lose their efficiency of photocatalysis due to mass transfer [7,8]. However, only a fraction of studies deal with stemming of mass transfer of immobilized TiO_2 or doped-TiO_2 photocatalyst films. The most promising approach to overcome this disadvantage is by immobilization of TiO_2 in the porous polymer matrix such as epoxides, the most important classes of compounds used in the coating industry. These epoxy composites provide thin-layer durable coatings having mechanical strength and good adhesion to a variety of substrates [9]. Antimicrobial epoxy based surface coatings of walls and floors can fight the nosocomial menace [10] in hospitals.

The antibacterial function of a TiO_2 photocatalyst is markedly enhanced even with weak UV light, such as fluorescent lamps and with the aid of either silver or copper, which is harmless to the human body [11]. TiO_2 nano-fillers improve mechanical properties like crack resistance, surface characteristics and can also contribute to the photostability of the host material. The photostability and photocatalytic activity of epoxy/nano-TiO_2 coatings under UV irradiation has been reported by Calza et al. [12]. While doping TiO_2 with silver can synergistically enhance photobactericidal acitivity of TiO_2, a considerable improvement in mechanical properties can also be achieved by introducing very low amount of nano-fillers into resin system [13]. In addition, photo-stability of epoxy resin can be improved by the presence of nano-TiO_2 by its UV absorption properties [14]. Thus, modification of polymers with TiO_2 and subsequent coupling with Ag^+/Ag NP enhance the photocatalytic and antimicrobial property of the material. Nanoparticles are generally introduced into epoxy matrix using various approaches like, in situ synthesis by reacting the precursors or physical dispersion of pretreated nano-fillers by mechanical stirring and subsequently processed by ultrasonication [15,16]. Successful dispersion of nanoparticles within the polymer matrix is determined by factors like particle size, particle modifications, specific surface area, particle load and the particle morphology.

Broadly there are two methods to impregnate a biocidal agent in order to achieve antibacterial polymeric materials. That is, either by introduction of aleaching biocidal agent into the polymer to form a composite or by covalent functionalization of the polymer with the pendent groups that confer antimicrobial activity. Such materials have displayed potent and broad spectrum antimicrobial activity [17]. The polycaprolactone-titania nanocomposites have been shown to decrease surface colonization of Escherichia coli and Staphylococcus aureus [18]. Similarly, introduction of (+)usnic acid, a natural antimicrobial agent into modified polyurethane prevented biofilm formation on the polymer surface by Staphylococcus aureus and Pseudomonas aeruginosa [19]. The poly(ethylene terephthalate) (PET) was surface functionalized with pyridinium groups possessing antibacterial properties, as shown by their effect on Escherichia coli [20]. Highly potent antibacterial activity toward both Gram-positive and Gram-negative bacteria was demonstrated by composites consisting of a cationic polymer matrix and embedded silver bromide nanoparticles [21].

There are very few empirical reports that quantitatively assess inhibition of biofilm formation on polymer surfaces by employing indicator dyes (crystal violet/fluorescent dye). Crystal violet (hexamethyl pararosaniline chloride) is such a dye, which binds proportionately to the peptidogly and can be a component of bacterial cell walls. It has been used by Kwasny and Opperman [22] to evaluate the amount of biofilm formed by staining the thick peptidoglycan layer of Gram-positive bacteria, the thin peptidoglycan layer of Gram-negative bacteria. In this study, anti-biofilm activity of polymeric surfaces was measured by protocol adoption as described by Kwasny and Opperman with minor modifications. The optical density of destaining solution after washing crystal violet adsorbed onto biofilm was measured with a multi-well plate spectrophotometer (using a 96 well titer plate). The color intensity of destaining solution after washing has been shown to be proportional to the quantity of biofilm formed. This method makes more practical high-throughput screening of polymer surfaces for their antibiofilm activity.

Metallic silver/TiO$_2$ and silver ion doped TiO$_2$ system in the form of films, deposition and its antibacterial performance in visible/UV light have been reported frequently [23–25]. To the best of our knowledge, there have been limited reports on the synthesis of polymers loaded with silver doped titania, for durable photobactericidal coatings that is compatible with many substrates to fight biofilms. In this work, composite materials suitable for coating was obtained by the addition of Ag-TiO$_2$ nanoparticles into epoxy resin system, with the aim to achieve "*in situ*" formation of silver species by photoreduction. The antibiofilm activity of this composite system is exhibited by the actions of photokilling and release of biocide (Ag$^+$/Ag^0) upon contact with aqueous environment.

2. Experimental Section

2.1. Preparation of Nanocrystalline TiO$_2$ and Ag-TiO$_2$

Ethanol 99.9%, Titanium(IV) butoxide, silver nitrate and acetic acid were of analytical grade and procured from Sigma Aldrich (Bangalore, India). About 1.5 wt% of Ag$^+$ ion doped nanocrystalline anatase TiO$_2$ was prepared by homogeneous hydrolysis of titanium butoxide-ethanolic solution using acetic acid-water as acid catalyst. The stoichiometric amount of AgNO$_3$ was dissolved in aqueous acetic acid and then added drop wise into the titania sol with stirring for 30 min at room temperature, and allowed to stand for two days at room temperature. Undoped TiO$_2$ gel was prepared by the same procedure without the addition of AgNO$_3$. All the gels were isochronally annealed initially at 100 °C for 2 h then at 500 °C for 4 h.

2.2. Nanocomposite Preparation and Coating

The commercial grade resins, Lapox® L-12 [liquid epoxy resin based on bisphenol-A, (4,4'-Isopropylidenediphenol, oligomeric reaction products with 1-chloro-2,3-epoxypropane)] and reactive diluent, Lapox® XR-19 (Diglycidyl ether of polypropylene glycol) were procured from Atul Ltd., Ahmedabad, India. Diethylenetriamine (DETA) as a curative agent from Sigma-Aldrich was employed. The low molecular weight epoxy Lapox® XR-19, was added as diluents to lower the viscosity of the base resin and improve the initial physical dispersion of TiO$_2$ in the epoxy. The nanocomposites were prepared as follows: (i) the resin mixture was prepared (resin + diluant); (ii) the resin solution was diluted with ethanol to further decrease the viscosity of the resin mixture at 1:5 ratio; (iii) different amount of TiO$_2$ or Ag-TiO$_2$ was mixed into the diluted resin mixture. Then, the mixtures were sonicated under water bath for 30 min and degassed under vacuum. The resin-to-curative ratio in the material preparation at 10% of resin mixture weight was added. The mixtures were spin coated into the 50 mm × 12 mm (outer dia × height) size Borosil® S-Line petri plate on flat bottom dish and allowed to dry at room temperature for 24 h. The coatings were postcured at 100 °C for 2 h. Six different material samples were coated—neat epoxy resin, undoped TiO_2/epoxy composite with 1 wt% loading and Ag-TiO_2/epoxy composite with 0.5, 1.0, 1.5 and 2.0 wt% loading Figure 1. The epoxy/Ag-TiO_2 composite turned pale brown indicating the formation of silver nanoparticles due to photoreduction. The coated substrates were sterilized by autoclaving at 121 °C, for 15 min before the start of experiments.

Figure 1. Assay petri dishes spin coated with neat epoxy, epoxy/TiO_2 and epoxy/Ag-TiO_2 composites.

2.3. Physicochemical Characterization

Powder X-ray diffraction (PXRD) measurements were recorded by Bruker D8 Advance (Bruker AXS Inc., Madison, WI, USA) X-ray diffractometer with Cu Kα radiation (1.5418 Å) at a 40 kV accelerating voltage and 30 mA. Raman measurements were performed with Renishaw Raman Microspectrometer (RM1000 System, Renishaw, Tokyo, Japan) of spectral resolution of 1 cm^{-1} and spatial resolution of ~2.5 nm (using 50X Objective and 514.5 nm laser line). Scanning electron microscopy (SEM) images were captured using a Philips XL30 CP microscope equipped with EDX (energy dispersive X-ray) (Philips, Eindhoven, The Netherlands). The Brunauer–Emmett–Teller (BET) surface area (calculated from nitrogen adsorption data) was measured on a Quantachrome NOVA 1000 system at −180 °C. UV-Vis diffuse reflectance spectra (DRS) were recorded using Analytik Jena Specord S600 spectrometer (Analytik Jena AG, Jena, Germany) (diffuse reflectance accessory with integrating sphere) by using $BaSO_4$ as a reference. All the above charecterizations were performed for the prepared nanocrystalline TiO_2 and Ag-TiO_2. The thermal property of composite materials was investigated by differential scanning calorimetry using Mettler-Toledo DSC823e (Mettler-Toledo AG, Schwerzenbach, Switzerland), and scans were performed at 5 °C/min for each composite under nitrogen flow and T_g value was extrapolated from the curves of second run.

2.4. Quantitative Determination of Biofilm

Bacteria used in this study were biofilm-proficient *S. aureus* ATCC 6538 and *E. coli* K-12 strains. Biofilm formation was measured under static condition by adopting quantitative crystal violet (CV) binding assay of Kwasny and Opperman with modifications [22]. In the current study, the flat inner surface of glass petri dish coated with prepared composites and resin was overlaid with 4 mL of sterile nutrient broth (composition is tabled in Supplementary Materials), so that the total area of the coating was covered. Then, 0.1 mL of logarithmic phase cultures of either *E. coli* or *S. aureus* grown over night to an optical density of *ca.* 0.1, at 595 nm in the appropriate growth media, were inoculated into sterile media in coated bottom plates prepared as above. Inoculated bottom plates were incubated in a bacteriological incubator at 37 °C under UV-A irradiation with intensity of 0.2 mW/cm^2 with λ_{max} around 365 nm (which is harmless to cause bacterial reduction), for different exposure durations. Later, the broth with planktonic cells was discarded by decantation. The plates were washed twice by gentle swirling with 2 mL of sterile phosphate-buffered saline to remove any non-adherent cells. Cells which remained adherent (biofilm mass) to the surface of polymer coated bottom plate were fixed by heating in a hot air oven at 60 °C for 60 min. Later plates were cooled to room temperature and stained with 1 mL of 0.06% (w/v) solution of crystal violet which was allowed to stand at room temperature for 5 min. Then plates were washed several times with phosphate-buffered saline to remove excess CV staining. Biofilm bound CV was eluted by vortexing with 1 mL of 30% acetic acid (destaining solution) for 10 min. The 0.2 mL aliquots of the wash solution with eluted crystal violet were transferred to 4 different wells of 96-well microtiter plates for the purpose of measuring the absorbance at 600 nm. Results were expressed as inhibition percentages of biofilm development. The percent inhibition of biofilm growth produced by each nanocomposite surface was calculated with the formula,

$$\left\{ 1 - \left[CV\ OD_{600}\ (\text{composite}) \middle/ \text{average}\ CV\ OD_{600}\ (\text{negative control}) \right] \right\} \times 100 \qquad (1)$$

where CV OD$_{600}$ is OD of crystal violet destaining solution obtained at λ_{max} 600 nm. The results are presented as the average of four individual replicates. To check the binding affinity of CV to the prepared composites and neat epoxy, a similar assay with 48 h of UV exposure was conducted as above with the plain broth which was not inoculated with bacteria. The OD of destaining solution when measured was found to be insignificant to interfere with the experimental results. Then, the resulting silver concentrations in the same plain broth were also quantified by atomic absorption spectroscopy (AAS) analysis, released into the exposed media by the composites of different Ag-TiO$_2$ loadings. AAS analysis of released silver concentration was carried out with a 7700X instrumentation (Agilent, Santa Clara, CA, USA), using different standard concentrations.

The reduction in biofilm colonization on composite was also determined in terms of CFU (colony forming unit), by sonicating assayed composite plate with 5 mL PBS for 5 min to remove adherent bacteria. The PBS suspension of released cells was then diluted appropriately, and spread on nutrient agar plate. The bacterial CFUs per milliliter of PBS that formed upon the medium was determined after incubation for 48 h at 37 °C. The experiment was repeated two times under identical conditions along with negative control (neat epoxy). The biofilm log reduction values were determined as difference between Log$_{10}$ CFU/plate recovered from the treated plates and Log$_{10}$ CFU/plate recovered from control plate (neat epoxy). Each experiment was conducted with three replications for each composite plates and colonies were enumerated to obtain the log reduction.

3. Results and Discussion

3.1. Characterization of Materials

Sol-gel derived nanocrystalline TiO_2 were subjected to the XRD analysis to determine crystalline phase and crystallite size. Titania exists in three crystalline polymorphs–anatase, rutile and brookite forms. Among these, anatase titania has been shown to exhibit higher antimicrobial activity than the other two and thus pure anatase phase content is a desirable feature [26]. The PXRD of titanias synthesized in this work had the peaks characteristic of anatase phase Figure 2a. (JCPDS No. 21-1272). From the X-ray diffraction patterns, the size of anatase TiO_2 materials prepared were in the nanometric scale Table 1. The average crystallite size was determined from the (101) plane in the PXRD pattern using Scherer's formula. The calculated value of undoped TiO_2 had bigger crystallite size while Ag-doped TiO_2 showed a decrease in the crystallite size. A good correlation between the Raman and PXRD was also observed Figure 2b. The changes in the crystallite size of TiO_2 nanocrystals upon Ag-doping are closely correlated to the broadening and shifts of the Raman bands with decreasing particle size [27]. Similar observations were made for the titania sysnthesised in the present work. During annealing process, silver nitrate thermally decomposes into silver. Bigger ionic radii of Ag^+ (0.75 Å) compared to Ti^{4+} (0.605 Å) prevents it from entering the crystal lattice of anatase TiO_2 because of a high energy barrier. Thus, it gets distributed uniformly on the surface of TiO_2. However, the PXRD pattern of Ag-TiO_2 did not reveal any Ag or Ag-containing phases. This may be due to the low concentration of Ag incorporated which is below the detection limit of the PXRD analysis.

Doping with Ag^+ ion also resulted in increase in the BET surface area of TiO_2 (48 m^2/g), while that of undoped TiO_2 showed BET surface area of 27 m^2/g. Thus, large surface area to volume ratio of Ag-doped TiO_2 was advantageous for the release of Ag^+ ion. From the energy dispersive X-ray (EDS) analysis at two locations (see Figure 3a), done during the SEM confirms silver is dispersed uniformly in TiO_2 host. Figure 3b shows the changes in the absorbance of Ag-doped TiO_2 in comparison to undoped TiO_2 and Degussa P 25 titania. Ag doped TiO_2 (calcined in ambient air at 500 °C) was found to have higher visible absorbance. In contrast, pure TiO_2 prepared under similar experimental conditions, had its absorbance slightly shifted towards the visible region as compared to Degussa P25 titania (Figure 3b). The DRS spectra showed a characteristic absorption band at about 500 nm, due to the surface plasmon resonance of silver [28]. Using the different absorbance onsets, it was found that the Ag-TiO_2 had a bandgap of ~2.8 eV while both of the undoped titania samples had wider band gaps estimated at ~3.1 eV for TiO_2 and ~3.2 eV for the Degussa P25 TiO_2 sample. Similar observations from previous studies can be confirmed [29].

Table 1. Physio-chemical properties of nanofiller, T_g, weight of coated composite material and amount of silver ion released.

Composite type	Nanocrystalline-TiO₂		Epoxy-TiO₂ composite		Amount of Ag released by the composite (μg/mL) *
	Crystallite size (nm)	BET surface area (m²/g)	Glass transition temperature T_g (°C)	Weight of the coated composite (gm)	
Neat Epoxy	n/a	n/a	93	1.08	Nil
1.0 wt% Epoxy/TiO₂	36	27	90	0.99	Nil
0.5 wt% Epoxy/Ag-TiO₂	18	48	94	1.09	6.6
1.0 wt% Epoxy/Ag-TiO₂	18	48	97	0.95	10.2
1.5 wt% Epoxy/Ag-TiO₂	18	48	106	1.05	14.6
2.0 wt% Epoxy/Ag-TiO₂	18	48	97	1.10	16.8

* Concentration of silver in the exposure media as determined by Atomic Absorption Spectroscopy (AAS), after 48 h.

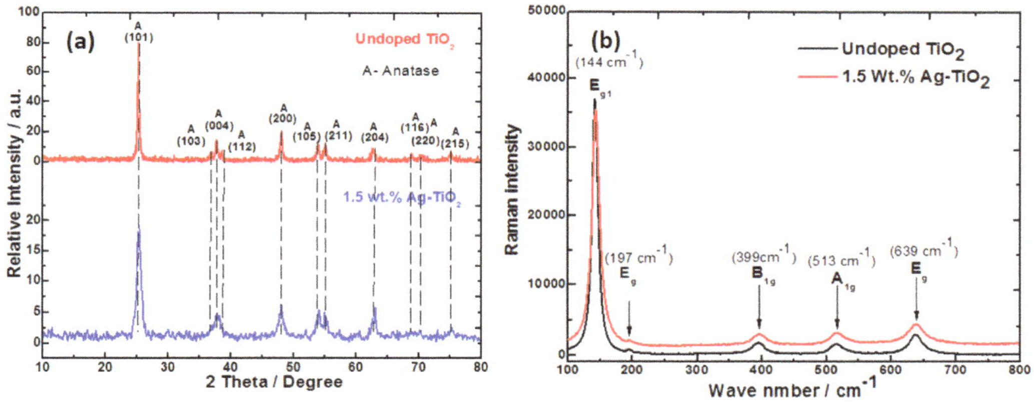

Figure 2. (**a**) Powder X-ray diffraction (XRD) and (**b**) Raman spectra of TiO₂ and Ag-TiO₂.

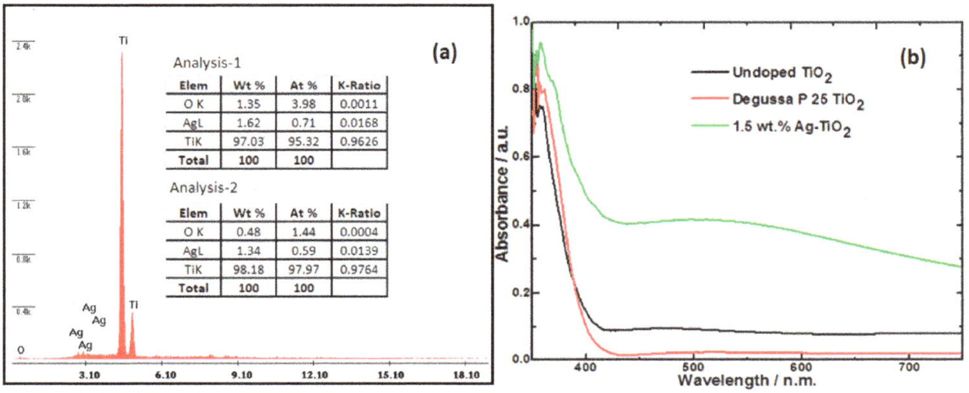

Figure 3. (**a**) Elemental analysis (EDS) of the silver doped TiO₂ showing the presence of Ti and Ag species; (**b**) UV-Vis diffuse reflectance spectra (DRS) of Ag-doped TiO₂, TiO₂ and Degussa P25 titania.

The homogeneous distribution of nano-filler in a polymer matrix has major influence on the composite performance. The morphology of synthesized titania nanoparticles and their dispersion in epoxy matrix were examined by SEM analysis Figure 4. The primary particle size of undoped and silver doped titania are different, varying from nanometer to micron size for the same magnification as seen in SEM micrographs Figure 4a,b. The undoped sample exhibited a nanostructure consisting of spherical clusters with a diameter of 50–500 nm, which are extensively agglomerated with an average crystallite size of 36 nm. However, silver doped titania showed bigger aggregates and smaller segregated particles consisting of primary anatase nanocrystals of 18 nm size (Figure 4b). Dispersion is an important factor in determining a nanocomposite's properties. Composites with the same weight percent (1 wt%) of nanofiller showed different degree of dispersion Figure 4c,d. The unmodified TiO_2 although thoroughly distributed in the matrix, yet particles agglomerated densely as shown in Figure 4c giving scattered hill lock like appearance on the surface of the composite. The size of these agglomerates varied from nanometers to micrometers. However, the $Ag-TiO_2$ particles Figure 4d, showed a lesser degree of agglomeration; interparticle distance are clearly visible between the TiO_2 particles. This indicates that the presence of silver enable good dispersion due to the interaction of oxidized silver ions with surface hydroxyl groups (titanol groups, Ti–OH) of TiO_2 and increase its wettability in apolar media like epoxy (hydrophobic polymer matrix). While Figure 4e shows the fractured surface of the composite, dispersion in the bulk is similar to distance between agglomerates as on surface. This suggests that the doped nano-fillers have better dispersion due to surface modifications, which improve the interactions between particles and polymer matrix. Use of reactive diluant also significantly reduced viscosity of epoxy resin during preparation and optimized the dispersion along with sonication.

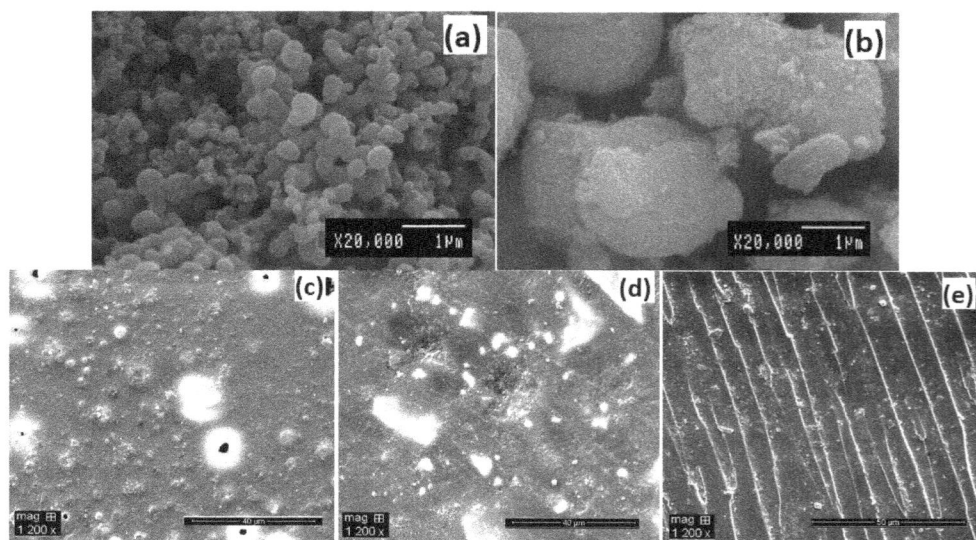

Figure 4. Scanning electron microscopic (SEM) characterization of (**a**) sol-gel synthesized TiO_2; (**b**) 1.5 wt% silver doped TiO_2; (**c**) 1 wt% epoxy/TiO_2 composite; (**d**) 1 wt% epoxy/$Ag-TiO_2$ composite; (**e**) Fractured surface of 1 wt% epoxy/$Ag-TiO_2$ composite.

The glass transition temperature (T_g) of the samples were determined from the tangents of DSC spectra as a function of temperature. The DSC curves of the neat epoxy and nanocomposites with 1 wt% of TiO_2 and Ag-TiO_2 nanofiller from the second run are shown in the Figure 5a. For thermosetting resin glass transition temperature (T_g), values can shift due to reasons like cross-linking density, intermolecular interaction and chain length. The addition of nanometer sized TiO_2 particles in epoxy resulted in increase in the T_g from 93 °C for neat epoxy to 97 °C at 1 wt% loading of Ag-TiO_2. Whereas, T_g of composite shifts to lower temperature with undoped TiO_2 (1 wt% loading) due to poor dispersion and agglomeration as evident in the SEM micrograph. Nanocomposites with Ag-TiO_2 exhibited maximum T_g value at 1.5 wt% loading (107 °C) (Figure 5b). A further increase in the nano-filler content to 2 wt% led to the drop in the T_g value, this is due to their easy agglomeration arising from van der Waals attraction between particles.

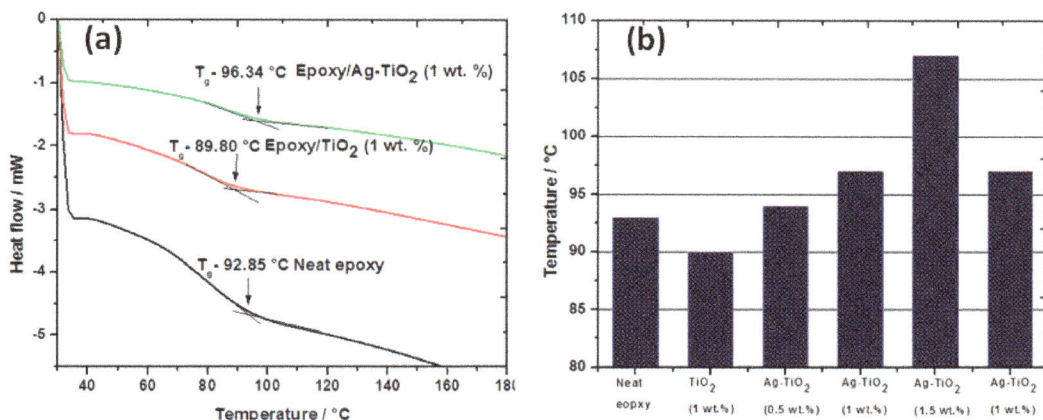

Figure 5. (**a**) DSC thermograms of neat epoxy and nanocomposites with 1 wt% of TiO_2 and Ag-TiO_2; (**b**) Variations in T_g values of neat resin and nanocomposites at different wt% of TiO_2/Ag-TiO_2 loading.

It can be seen from Figure 5b that the T_g value increases steadily then value drops; this corroborates with the trend observed by other investigators [13,30]. With our study, the degree of dispersion and nanofiller loading affected the shifts in T_g for epoxy/Ag-TiO_2 composites. The size, loading and dispersion state of the nanofillers are the factors that impact the glass-transition temperature. The T_g value increases due to polymer chain-filler (organic-inorganic interfacial contact) that are immobilized by cohesive interactions at the interface of nanofiller in the bulk of the material. On the other hand, higher loading of nanofiller or their agglomeration can result in mobile moieties within the matrix which significantly decrease the glass transition temperature. Very high T_g values are not achievable by room temperature curing agents, and the composites reported here can find their applications at temperature conditions below their T_g. These synthesized epoxy composites may be cross linked by means of any conventional hardener at room temperature, without the decomposition of incorporated biocides.

3.2. Antibiofilm Activity on the TiO₂ and Ag-TiO₂ Nanocomposite Coatings

Antibacterial epoxy coatings for antibiofilm properties were tested against *S. aureus* and *E. coli* under static conditions in glass petri dish with UV-A irradiation, on the surfaces of TiO₂ and Ag-TiO₂ composites (both with 1 wt% loading). Both *S. aureus* and *E. coli* were able to form biofilm on neat epoxy resin surface (negative control) and composites, *i.e.*, biofilm formation was independent of the underlying composite substrates. In the absence of TiO₂, epoxy resin showed higher growth of biofilm than that of epoxy/TiO₂ composite. Anti-biofilm activity appeared to increase significantly for Ag-TiO₂ composite.

The biofilm inhibition by composites does not seem to be restricted to specific strains or growth conditions; *E. coli* and *S. aureus* varied in their ability to produce biofilm on the surface of the composites as shown in Figure 6. In all assays, the amount of crystal violet eluted from *E. coli* biofilms was lower than that of *S. aureus* biofilms, because *E. coli*, being a Gram negative organism binds lesser dye than Gram positive organisms like *S. aureus*. The OD$_{600}$ of CV eluates from both biofilms was in the range of 0.121 to 2.8. Among the bacterial pathogens, *E. coli* was more susceptible for biofilm inhibition than *S. aureus* on these surfaces.

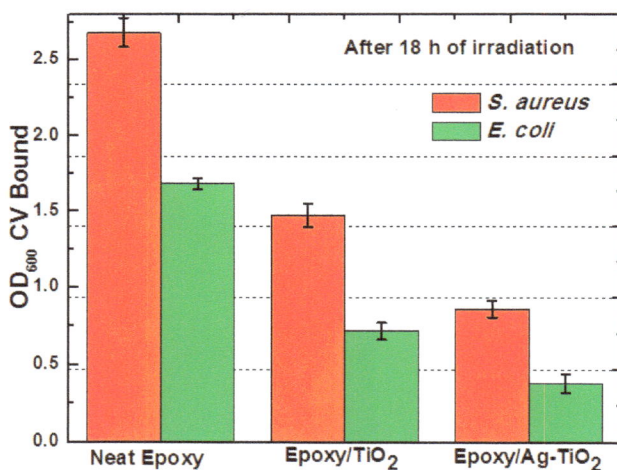

Figure 6. Spectrophotometric analysis (OD$_{600}$) of solubilized crystal violet of *E. coli* and *S. aureus* biofilm at 18 h irradiation time on the surfaces of TiO₂ and Ag-TiO₂ composite with similar loading (1 wt%).

To confirm the activity of TiO₂/Ag-TiO₂ on the surface of nanocomposite for the photokilling, we conducted the experiments under both dark and irradiated conditions as shown in Figure 7, and we found that higher inhibition of biofilm under irradiated conditions as shown in Figure 7b. The Ag-TiO₂ composite (1 wt%) showed 24% and TiO₂ composite (1 wt%) showed 6% biofilm inhibition of *E. coli* after 18 h of incubation in the dark as shown in Figure 7a. For the same conditions with UV irradiation *E. coli* biofilm showed 56% inhibition for epoxy/TiO₂ and 77% inhibition for epoxy/Ag-TiO₂, while that of *S. aureus* biofilm showed 43% and 67% ihibition, for epoxy/TiO₂ and epoxy/Ag-TiO₂ composites respectively. It is, therefore, the bactericidal activity of silver on biofilm that is rendered more likely in the absence of photokilling by Ag-TiO₂ with the dark experiment data. However, enhanced antibiofilm response of Ag-TiO₂ composite under UV irradiation can be attributed to the silver surface plasmon band favoring UV light absorption along with nanometer sized silver particles which exhibited a striking degree of synergy. The antibacterial feature was diminished for epoxy/TiO₂ composite in the dark experiment. However, the bare TiO₂ particles which are non-photo-activated on the surface also supported minor antibacterial activity, even in the dark. This is due to direct attack of cells upon contact with TiO₂ nanoparticles which disrupt the integrity of the bacterial membrane [31,32]. This is also in agreement with reported experimental findings by Gogniat *et al.* [33] who also showed a loss of bacterial culturability after contact with TiO₂ nanoparticles even in the dark. These data show that the nature of epoxy resin makes it suitable host for dispersion of photocatalyst like TiO₂ for bacteriacidal activity.

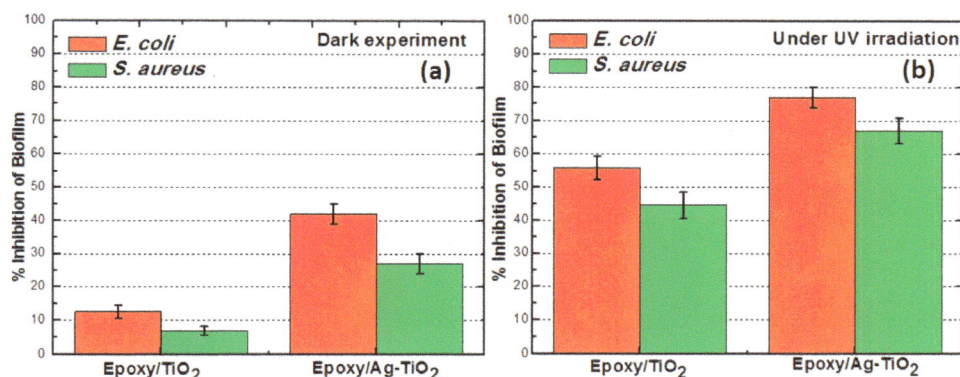

Figure 7. Mean values of quadruplicate experiments showing percent inhibition of *E. coli* and *S. aureus* bio-film formation on epoxy/TiO₂ and epoxy/Ag-TiO₂ composite surfaces calculated relative to the neat epoxy (negative control), under (**a**) dark and (**b**) UV irradiated conditions.

The release of the antimicrobial species (Ag⁺, Ag⁰ and ROS) from a composite occurs due to the interaction of the diffused water molecules with TiO₂ and dispersed silver within the matrix during UV exposure; upon submerging it in the culture media [34,35]. Silver ions resident within the metal oxide nanofiller can diffuse to the surface of the epoxy matrix. The leaching of Ag⁺ ions was confirmed by AAS analysis of the bacterial media from blank experiments (without inoculums as explained in the experimental section). The Ag⁺ ion concentration of the same media was determined by atomic absorption spectrophotometer (AAS), which strongly suggests Ag⁺/Ag⁰ are associated noncovalently with cross-linked polymeric host and has leached to aqueous medium. By AAS analysis, the silver concentration (Ag⁺/Ag⁰) in the exposed media for the different epoxy/Ag-TiO₂ composite, showed a nonlinear increase that approached a maximum for the composite with 2.0 wt% of Ag-TiO₂ loading Table 1.

The valence band "electrons" can be excited to the conduction band (e_{cb}^-), leaving positive "holes" in the valence band (h_{vb}^+) to form an e⁻/h⁺ couple that react with aqueous environment and oxygen, to generate reactive oxygen speces (ROS) such as $OH^{\cdot-}$, $HO_2^{\cdot-}$ and $O_2^{\cdot-}$, which are responsible for the mechanistic photo-biocidal activity [36,37].The photoexcitation of non-leachably associated TiO_2 occurs when it absorbs light equal to or greater than band-gap energy near-ultraviolet light region. While Ag NP and Ag^+ could act as efficient electron scavengers, and significantly enhanced the visible light responsiveness of TiO_2 to generate more oxygen free radicals by improving the quantum efficiency of a charge pair generated [35]. At the same time, these oxygen species can reduce Ag^+ ions to form Ag nanoparticles. The smaller Ag^+ ions can easily penetrate the cell wall and thus can hasten antimicrobial activity.

The attack of Ag^+ on disulfide or sulfhydryl (thiol) groups present in the membrane protein result in formation of stable S–Ag bond with –SH groups thus inhibiting enzyme-catalyzed reactions and the electron transport chain that are necessary for biofilm formation [38]. We speculate that the outer membrane of the bacterial cell is attacked by photocatalytic oxidation enabling the antimicrobial metal ions/particles to diffuse to interior of the cell thus becoming much more lethal to the bacterium. Thus, capability of photoactiveTiO_2 and leachable silver in destabilizing the biofilm matrix is enhanced by synergistic approach.

3.3. Effect of Exposure Duration on Formation of S. Aureus and E. Coli Biofilms

Figure 8 shows OD_{600} values of eluted dye solution by *E. coli* and *S. aureus* for different duration of exposure (6 h, 9 h, 12 h, 15 h, 18 h, 20 h, 22 h and 24 h) of neat epoxy, epoxy/TiO_2 (1 wt%) and epoxy/Ag-TiO_2 (1 wt%). The biofilm ODs presented are averages of four independent experiments. Time course studies showed bactericidal ability of prepared composite surface up on contact and effectiveness in restraining bacterial biofilm formation. *S. aureus* biofilm formation response to time increased gradually, but it declined over a longer incubation period. It is plausible that this is due to biosorption of minerals and metals by microbial biofilms from the environment with which they are in contact [39,40]. When higher levels of silver is reached or with chronic exposure, it should be possible to limit the ability of the biofilm biosorption capacity, silver would then inhibit biofilm formation during prolonged exposure.

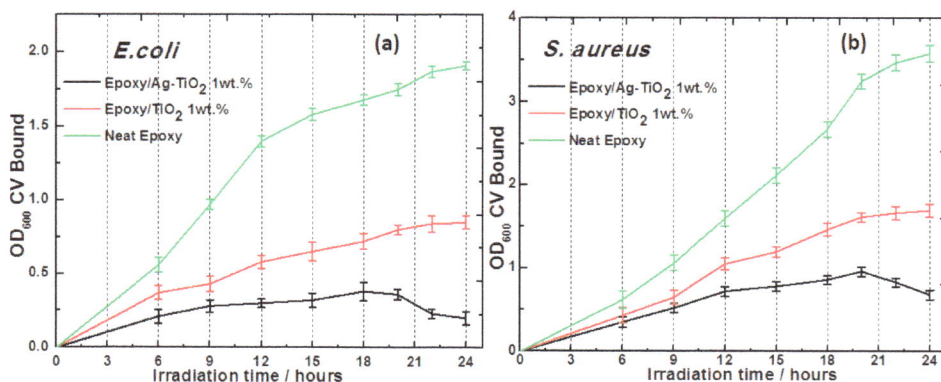

Figure 8. Growth curve for biofilm formation on neat resin, epoxy/TiO_2 and epoxy/Ag-TiO_2 composite of (**a**) *E. coli* and (**b**) *S. aureus*.

3.4. Effect of Ag-TiO$_2$ Loading on Biofilm Inhibition

The results showed that biofilm formation was highly inhibited in a dose dependent manner as shown in Figure 9. Increasing the load of Ag-TiO$_2$ resulted in shorter inhibition time *i.e.*, antibiofilm activity of composite is directly proportional to Ag-TiO$_2$ loading. Exposure of the composite with 1.5 wt% Ag-TiO$_2$ for 24 h. resulted in a inhibition of 100% (as per crystal violet binding assay) of both *E. coli* and *S. aureus*. The higher activity of these composites against *E. coli* a Gram-negative bacterium is attributed to its thinner peptidoglycan cell wall compared to *S. aureus* a Gram-positive bacterium. Complete inhibition of biofilm was achieved with 24 h of irradiation time with composite of Ag-TiO$_2$ with 1.5 wt% loading, in case of both *E. coli* and *S. aureus* (see Figure 9a,b). The antibacterial activity could also have effect on planktonic bacteria due to silver that has diffused to media from the matrix. The bactericidal efficacy of these composite is through the diffusion of photogenerated ROS and Ag$^+$ particles (acting as a leaching biocide) to the surface from the bulk of the polymer where such species/particles attack proteins and membrane lipids in bacterial cell wall. The driving force for silver particle diffusion is determined by a concentration gradient, which forms between the bulk of the composite material and the surface. The diffusion behavior depends on several factors including the structure of the material, environmental osmolarity and temperature.

We have quantified the silver release characteristics at 37 °C for the composites loaded with the 0.5 wt% to 2.0 wt% Ag-TiO$_2$ filler Table 1. And observed that non linear increase in the release of silver on increase of Ag-TiO$_2$ loading. The total released silver from the coatings was 6.6 to 16.8 µg/mL (16.8 ppm) after 48 h by epoxy/Ag-TiO$_2$ composites in the culture media without inoculum. From this observation it can be concluded that all the Ag-TiO$_2$ containing composites can have antibacterial activity even in the dark due to release of silver. However, presence of UV light will hasten the bactericidal activity of the composite due to photogeneration of ROS. Similar observations were made by Akhavan and Ghaderi [41] who investigated bactericidal activity of the anatase-TiO$_2$, the Ag thin film and the Ag-TiO$_2$/anatase-TiO$_2$ nanocomposite thin film against *E. coli* at dark and under UV exposure. In addition, they found superior antibacterial activity of Ag-TiO$_2$/anatase-TiO$_2$ nanocomposite thin film under the UV irradiation due its photocatalytic capability when compared to non-photocatalytic bare Ag and TiO$_2$ films and the silver ions released by Ag-TiO$_2$/anatase-TiO$_2$ nanocomposite thin film became saturated after 20 days at ~2 nM/mL. It is also possible to regulate the release of silver to the desired concentration by varying the nano-filler load incorporated into polymer composites and by tuning Ag-TiO$_2$ structure/composition during the sol-gel incorporation process. Antibiofilm activity of these composite remained unchanged at least for 5–6 cycles when we challenged during experiment through replications, this is due to continuous and uniform diffusion of the antimicrobial agents (ROS and silver species).

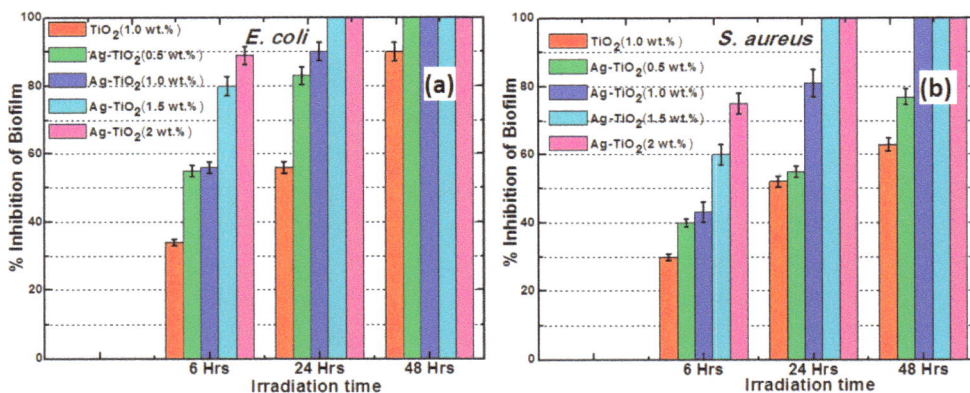

Figure 9. Biofilm inhibitory effect of Ag-TiO2 loading (dose response) after 6, 24 and 48 h of irradiation on (**a**) *E. coli* and (**b**) *S. aureus*.

3.5. Quantitative Comparisions

There is no general consensus evolved for the comparison of efficiency of antibacterial activity of polymers surfaces between the research groups. However, most studies on antibacterial activity are interpreted by the number of surviving colony forming unit CFU/mL^{-1} or per unit area. Kubacka *et al.* [42], studied the antibacterial effect of isotactic polypropylene (iPP) polymeric matrix incorporated with anatase-TiO2 against *Pseudomonas aeruginosa* (Gram negative) and *Enterococcus faecalis* (Gram positive). They reported a maximum reduction by *ca.* 8–9 log in 30 min in case of *P. aeruginosa*. Francolini *et al.* [19] evaluated the effect of (+)-usnic acid incorporated into modified polyurethane surfaces on the biofilm forming ability of *S. aureus*. After three days postinoculation, they found culturable biofilm cell concentration of *S. aureus* on the untreated polymer was 7.3 log_{10} CFU/cm^2 compared to 0.9 log_{10} CFU/cm^2 on the (+)-usnic acid-containing polymer. Cen *et al.* [20] introduced pyridinium groups at 15 $nmol/cm^2$ on the surface of poly(ethylene terephthalate) (PET) film and demonstrated its bactericidal effect against *Escherichia coli*. Jansen *et al.* introduced silver ions by plasma-induced grafting onto polyurethane films which was found to reduce adherent viable bacteria from initial 10^4 $cells/cm^2$ to zero within 48 h [43]. Jiang *et al.* [44] coated silver on silicon rubber substrates and showed decline in number of *L. monocytogenes* cells post 6 h. After 12 h, there was a reduction of over 2-log_{10} CFU/chip, and no viable bacteria were detected on both types of silver-coated SR after 18 and 24 h. Sambhy *et al.* [21] demonstrated antibacterial activity of composites consisting of poly(4-vinyl-N-hexylpyridinium bromide) (NPVP) embedded with silver bromide nanoparticles. They observed no biofilm formation on 1:1 AgBr/21% NPVP-coated surfaces after 72 h when incubated for 24–72 h with *P. aeruginosa* suspension (10^7 CFU/mL) in LB broth. Pant *et al.* [45] have demonstrated the ability to eliminate up to 99.9% of pathogenic bacteria on the surface of siloxane epoxy system containing quaternary ammonium moieties. In another work involving epoxy system, Perk *et al.* [46] observed fungicide, carbendazim supported on poly (ethylene-*co*-vinyl alcohol) and epoxy resin coating showed the antifungal activity contingent upon release from their polymer supports.

Coatings and thin films based on titania photoctalysts (Ag$^+$-doped TiO$_2$/Ag-TiO$_2$/TiO$_2$) that kills microbes under UV and visible light illumination, also have been actively investigated in recent years. Studies by Necula et al. [47], with TiO$_2$-Ag composite coating prepared by plasma electrolytic oxidation on implantable titanium substrate, showed the ability to completely kill methicillin-resistant S. aureus (MRSA) within 24 h. In yet another investigation by Necula et al. [25], they examined the ion release and antibacterial activity of porous TiO$_2$-Ag coating on biomedical alloy disk. Each evaluated samples could release 20.82 and 127.75 μg of Ag$^+$ per disk and showed markedly enhanced killing of the MRSA inoculums with 98% and >99.75% respectively within 24 h of incubation, while their silver free counterpart sample allowed the bacteria to grow up to 1000-fold. The non-cumulative release of silver ions of 0.4 ppm, 0.26 ppm and 0.005 ppm for 1 h, 24 h and 7 days respectively after immersion in water, from nanometer scale Ag-TiO$_2$ composite film was demonstrated by Yu et al. [34] and they also reported that 0.4 ppm released silver from Ag-TiO$_2$ composite film is sufficient to cause almost 100% killing of E. coli when exposed to UV for 1 h. Studies by Jamuna-Thevi et al. [48], reported nanostuctured Ag$^+$ doped TiO$_2$ coatings deposited by RF magnetron on stainless steel, with overall Ag$^+$ ion release measured between 0.45 and 122 ppb. They also noted that at least 95 ppb Ag$^+$ ion released in buffered saline was sufficient for 99.9% of reduction against S. aureus after 24 h of incubation. Biological activity of silver-incorporated bioactive glass studies conducted by Balamurugan et al. [49] assessed in vitro antibacterial bioactive glass system elicited a rapid bactericidal action. Antimicrobial efficacy of these silver-incorporated bioglass suspension at 1 mg/mL for E. coli was estimated to be >99% killing, and the amount of Ag$^+$ released from silver-incorporated glass was up to 0.04 mM after 24 h. In yet another study involving silver ions release by Liu et al. [35], the amount of silver released form the mesoporous TiO$_2$ and Ag/TiO$_2$ composites was measured to be 1.6×10^{-8} mol after 20 days. The photo-bactericidal activity on composite films was extremely high and displayed bactericidal activity even in the dark; they further reported that the survival rate was only 9.2% in the dark, and the E. coli cells were totally killed in UV light. Sun et al. [50] reported killing of bacteria on Ag-TiO$_2$ thin film, even in the absence of UV irradiation against S. aureus and E. coli with significant antibacterial rate about 91% and 99% after 24 h respectively due to release of silver, and the concentration of silver ions released from the Ag-TiO$_2$ film was 0.118 μg/mL during 192 h. Akhavan [51], reported that a concentration of 2.8 to 2.5 nM/mL completely killed 10^7 CFU/mL E. coli with visible light response photocatalytic Ag-TiO$_2$/Ag/a-TiO$_2$ material in 110 min. However, in most of the cases reports are based on planktonic studies and the release of silver is dependent upon the method employed for coating, thickness, conditions for gradient formation and silver source used. Nevertheless, release of silver ions frombare Ag/TiO$_2$ composite layers reported above, obtained by methods viz., impregnation, deposition and nano-coatings gradually diminish over the time.

Bacterial biofilms are often more difficult to eradicate unlike planktonic cells. Until now, there have been very few reports that shown to resist biofilm formation on titania based polymer-nanocomposites. In one such study, Kubacka et al. [52] have demonstrated photocatalysis using ethylene-vinyl alcohol copolymer (EVOH) embedded with Ag-TiO$_2$ nanoparticles (ca. 10^{-2} wt%) that showed outstanding resistance to biofilm formation by bacteria and yeast, upon ultraviolet (UV) light activation. In the present study, although the release kinetics of silver was not established but comparing to above studies which established the antimicribial threshold concentration of silver and efficacy of killing with different bare Ag-TiO$_2$ (Ag/Ag-TiO$_2$ nanofilms), the polymer composite system reported here which released 6.4 to 16.8 μg/mL of silver seems adequate [53], when the overall biocidal ability (to prevent bacterial attachment) of the composite during 48 h period in combination with radical-mediated photocatalytic action. Practically, the added strengths of the polymer-based Ag-TiO$_2$ nanocomposite coatings as compared to bare TiO$_2$/Ag-TiO$_2$ coatings are its wear stability, flexibility, permeability and optical properties.

But the main objective of the disinfection technology in ensuring microbiological safety is to; set a standard for achieving a required logarithm of reduction of the microbial consortia. The microbial cells, which are not inactivated by the antimicrobial coatings adhering onto the testing surface over the different irradiation time, were able to grow on the agar plates. Quantifying their reduction in number (for quantitative assessment) of surviving CFU on a bactericidal surface compared with a non-bactericidal (neat epoxy) surface revealed reduction of microbial cells. In the present study, epoxy/Ag-TiO$_2$ with 1.0 wt% loading was found to cause a reduction of CFU on agar plates by approximately 6-log in case of E. coli and the same effected ca. 4-log reduction in case of S. aureus after 48 h of incubation, while epoxy/TiO$_2$ with 1.0 wt% loading exhibited lesser inhibition of biofilm formation, see Table 2.

There was an initial slower decrease in bacterial load by all the composites, i.e., below 1-log reduction observed up to 18 h exposure followed by a rapid microbial decrease up to 6-log in 48 h for both 1.0 wt% of TiO$_2$ and Ag-TiO$_2$ loaded epoxy composites. Incomplete inhibition of biofilm formation was observed with lesser Ag-TiO$_2$ loading, but complete inhibition of both E. coli and S. aureus was possible for composites with above 1.5 wt% of Ag-TiO$_2$ after 24 h with UV irradiation. Strikingly, for the composite coating with 2.0 wt% epoxy/Ag-TiO$_2$ showed highest antibiofilm effectiveness with 1-log reduction in 18 h, i.e., the shortest period with maximum inhibition. In addition, after 48 h of irradiation against both S. aureus and E. coli with very few surviving CFUs and complete inhibition (biofilm formation) and 7-log reduction was observed, relative to that in control plates as shown in Table 2. However, the present study results take into consideration only biofilm phase inhibition, and the obtained concentrations of range 6.4–16.8 μg/mL (ppm) Ag$^+$ is very high (many times above minimum biocidal concentration levels) to radically prevent microbial cell viability. The polymer-based nanocomposite reported here obtained by dispersion of the Ag-TiO$_2$ nanoparticles into epoxy manifest a real potential as photobiocidal coatings in a wide variety of settings that prevents biofilm formation by a wide range of Gram-positive and Gram-negative bacteria.

Table 2. Different nanocomposite materials and their antibiofilm efficacy for 18 and 48 h irradiation time.

Composite type	E. coli (G⁻ᵛᵉ)		S. aureus (G⁺ᵛᵉ)	
	% Inhibition [a]	Log₁₀ Reduction [b]	% Inhibition [a]	Log₁₀ Reduction [b]
% Biofilm inhibition and Log CFU reduction after 18 h				
1wt% Epoxy/TiO₂	57.2 (±1.5)	<1.0 (±0.03)	46.0 (±1.4)	<1.0 (±0.02)
1wt% Epoxy/AgTiO₂	77.0 (±1.4)	<1.0 (±0.02)	68.5 (±2.0)	<1.0 (±0.02)
2wt% Epoxy/AgTiO₂	90.0 (±1.3)	1.0 (±0.2)	90.0 (±1.4)	1.0 (±0.03)
% Biofilm inhibition and Log CFU reduction after 48 h				
1wt% Epoxy/TiO₂	90.0 (±1.0)	1.0 (±0.2)	63 (±0.9)	1.0 (±0.2)
1wt% Epoxy/AgTiO₂	100	6.0 (±0.18)	99.9 (±0.1)	4.0 (±0.11)
2wt% Epoxy/AgTiO₂	100	7.0 (±0.19)	100	7.0 (±0.2)

[a] Percent reduction in biofilm formation as determined by Crystal Violet assay; [b] Mean value ± SD for the group Log₁₀ reduction in CFU/plate.

4. Conclusions

The investigation relates the preparation of antibiofilm composite coatings containing both photocatalytic non-leaching Ag-doped TiO₂ and leaching silver biocide for production of potent oxidants (ROS) and silver species at the surface. The antimicrobial activity of these composite surfaces was quantified based on the inhibition of biofilm formation using crystal violet assay, which can be adopted more conveniently in high-throughput experiments. These antimicrobial materials are capable of killing microorganisms upon contact by inhibiting the biofilm formation in the aqueous environments. Both epoxy/TiO₂ and epoxy/Ag-TiO₂ nanocomposites exposed to UV irradiation exhibited antibiofilm activity against *S. aureus* (Gram-positive) and *E. coli* (Gram-negative). Although the optimal antimicrobial conditions remain to be fully established, the results highlight a better antibiofilm activity of Epoxy/Ag-TiO₂ compared to Epoxy/TiO₂. The role of different silver species could be that Ag⁺ as an active species found to enhance the catalytic activity, in contrast, Ag⁰ species showing strong antibacterial activity. This material may find potential applications in designing self-disinfecting surfaces, especially for hospitals and food industries where hygiene is a high priority.

Acknowledgements

The authors gratefully acknowledge the support of Rastriya Shikshana Samithi Trust, Bangalore. The authors would also like to acknowledge Department of Materials Engineering, IISc, Bangalore, India, for the help in carrying out the XRD, Raman and SEM analysis. In addition, we acknowledge Aravind K. of Intelli Biotechnologies, Bangalore for helping in microbiological assays.

20

Author Contributions

S.S.M. prepared and charecterised the materials, performed the experiments, analyzed the data and designed the structure of manuscript. N.K. gave technical advice and reviewed the manuscript.

Conflicts of Interest

The authors declare no conflict of interest.

References

1. Del Pozo, J.L.; Patel, R. The challenge of treating biofilm-associated bacterial infections. *Clin. Pharmacol. Ther.* **2007**, *82*, 204–209.
2. Stewart, P.S.; Costerton, J.W. Antibiotic resistance of bacteria in biofilms. *Lancet* **2001**, *358*, 135–138.
3. Gupta, S.M.; Tripathi, M. A review of TiO_2 nanoparticles. *Chin. Sci. Bull.* **2011**, *56*, 1639–1657.
4. Daghrir, R.; Drogui, P.; Robert, D. Modified TiO_2 for environmental photocatalytic applications: A review. *Ind. Eng. Chem. Res.* **2013**, *52*, 3581–3599.
5. Keleher, J.; Bashant, J.; Heldt, N.; Johnson, L.; Li, Y. Photocatalytic preparation of silver-coated TiO_2 particles for antibacterial applications. *World J. Microbiol. Biotechnol.* **2002**, *18*, 133–139.
6. Nakano, R.; Hara, M.; Ishiguro, H.; Yao, Y.; Ochiai, T.; Nakata, K.; Murakami, T.; Kajioka, J.; Sunada, K.; Hashimoto, K.; *et al.* Broad spectrum microbicidal activity of photocatalysis by TiO_2. *Catalysts* **2013**, *3*, 310–323.
7. McMurray, T.A.; Byrne, J.A.; Dunlop, P.S.M.; Winkelman, J.G.M.; Eggins, B.R.; McAdams, E.T. Intrinsic kinetics of photocatalytic oxidation of formic and oxalic acid on immobilised TiO_2 films. *Appl. Catal. A General.* **2004**, *262*, 105–110.
8. Chen, D.; Li, F.; Ray, A.K. External and internal mass transfer effect on photocatalytic degradation. *Catal. Today* **2001**, *66*, 475–485.
9. Guo, Z.; Pereira, T.; Choi, O.; Wang, Y.; Hahn, H.T. Surface functionalized alumina nanoparticle filled polymeric nanocomposites with enhanced mechanical properties. *J. Mater. Chem.* **2006**, *16*, 2800–2808.
10. McIntosh, R.H. Self-sanitizing epoxy resins and preparation thereof. US Patent 4,647,601A, 1987.
11. Sunada, K.; Watanabe, T.; Hashimoto, K. Bactericidal activity of copper-deposited TiO_2 thin film under weak UV light illumination. *Environ. Sci. Technol.* **2003**, *37*, 4785–4789.
12. Calza, L.R.; Sangermano, M. Investigations of photocatalytic acitivies of different photosensitive semiconductors dispersed into epoxy matrix. *Appl. Catal. B Environ.* **2011**, *106*, 657–663.
13. Chatterjee, A.; Islam, M.S. Fabrication and characterization of TiO_2-epoxy nanocomposite. *Mater. Sci. Eng. B.* **2008**, *487*, 574–585.
14. Preuss, H.P. *Pigments in Paint*; Noyes Data Corp.: Park Ridge, NJ, USA, 1974.
15. Schmidt, H. Sol-gel derived nanoparticles as inorganic phases polymer-type matrices. *Macromol. Symp.* **2000**, *159*, 43–55.

16. Bittmann, B.; Haupert, F.; Schlarb, A.K. Preparation of TiO$_2$ epoxy nanocomposites by ultrasonic dispersion and resulting properties. *J. Appl. Polym. Sci.* **2012**, *124*, 1906–1911.

17. Kugel, A.; Stafslie, S.; Chisholm, B.J. Antimicrobial coatings produced by "tethering" biocides to the coating matrix: A comprehensive review. *Prog. Org. Coat.* **2011**, *72*, 222–252.

18. Muñoz-Bonilla, A.; Cerrada, M.; Fernández-García, M.; Kubacka, A.; Ferrer, M.; Fernández-García, M. Biodegradable polycaprolactone-titania nanocomposites: Preparation, characterization and antimicrobial properties. *Int. J. Mol. Sci.* **2013**, *14*, 9249–9266.

19. Francolini, I.; Norris, P.; Piozzi, A.; Donelli, G.; Stoodley, P. Usnic acid, a natural antimicrobial agent able to inhibit bacterial biofilm formation on polymer surfaces. *Antimicrob. Agents Chemother.* **2004**, *48*, 4360–4365.

20. Cen, L.; Neoh, K.G.; Kang, E.T. Surface functionalization technique for conferring antibacterial properties to polymeric and cellulosic surfaces. *Langmuir.* **2003**, *19*, 10295–10303.

21. Sambhy, V.; MacBride, M.M.; Peterson, B.R.; Sen, A. Silver bromide nanoparticle/polymer composites: Dual action tunable antimicrobial materials. *J. Am. Chem. Soc.* **2006**, *128*, 9798–9808.

22. Kwasny, S.M.; Opperman, T.J. Static biofilm cultures of Gram-positive pathogens grown in a microtiter format used for anti-biofilm drug discovery. *Curr. Protoc. Pharmacol.* **2010**, *50*, doi:10.1002/0471141755.ph13a08s50.

23. Zhang, Q.; Ye, J.; Tian, P.; Lu, X.; Lin, Y.; Zhao, Q.; Ning, G. Ag/TiO$_2$ and Ag/SiO$_2$ composite spheres: Synthesis, characterization and antibacterial properties. *RSC Adv.* **2013**, *3*, 9739–9744.

24. Sornsanit, K.; Horprathum, M.; Chananonnawathorn, C.; Eiamchai, P.; Limwichean, S.; Aiempanakit, K.; Kaewkhao, J. Fabrication and characterization of antibacterial Ag-TiO$_2$ thin films prepared by DC magnetron Co-sputtering technique. *Adv. Mater. Res.* **2013**, *770*, 221–224.

25. Necula, B.S.; van Leeuwen, J.P.T.M.; Fratila-Apachitei, E.L.; Zaat, S.A.J.; Apachitei, I.; Duszczyk, J. *In vitro* cytotoxicity evaluation of porous TiO$_2$-Ag antibacterial coatings for human fetal osteoblasts. *Acta. Biomater.* **2012**, *8*, 4191–4197.

26. Fu, G.; Vary, P.S.; Lin, C. Anatase TiO$_2$ nanocomposites for antimicrobial coatings. *J. Phys. Chem. B.* **2005**, *109*, 8889–8898.

27. Choi, H.C.; Jung, Y.M.; Kim, S.B. Size effects in the Raman spectra of TiO$_2$ nanoparticles. *Vib. Spectrosc.* **2005**, *37*, 33–38.

28. Seery, M.K.; George, R.; Floris, P.; Pillai, S.C. Silver doped titanium dioxide nanomaterials for enhanced visible light photocatalysis. *J. Photochem. Photobiol. A Chem.* **2007**, *189*, 258–263.

29. Jiang, Z.; Ouyang, Q.; Peng, B.; Zhang, Y.; Zan, L. Ag size-dependent visible-light-responsive photoactivity of Ag-TiO$_2$ nanostructure based on surface plasmon resonance. *J. Mater. Chem. A* **2014**, *2*, 19861–19866.

30. Ash, B.J.; Schadler, L.S.; Siegel, R.W. Glass transition behavior of alumina/polymethylmethacrylate nanocomposites. *Mater. Lett.* **2002**, *5*, 83–87.

31. Fujishima, A.; Rao, T.N.; Tryk, D.A. Titanium dioxide photocatalysis. *J. Photochem. Photobiol. A Photochem. Rev.* **2000**, *29*, 1–21.

32. Verdier, T.; Coutand, M.; Bertron, A.; Roques, C. Antibacterial activity of TiO$_2$ photocatalyst alone or in coatings on *E. coli*: The influence of methodological aspects. *Coatings* **2014**, *4*, 670–686.

33. Gogniat, G.; Thyssen, M.; Denis, M.; Pulgarin, C.; Dukan, S. The bactericidal effect of TiO2 photocatalysis involves adsorption onto catalyst and the loss of membrane integrity. *FEMS Microbiol. Lett.* **2006**, *258*, 18–24.

34. Yu, B.; Leung, K.M.; Guo, Q.; Lau, W.M.; Yang, J. Synthesis of Ag-TiO2 composite nano thin film for antimicrobial application. *Nanotechnology* **2011**, *22*, doi:10.1088/0957-4484/22/11/115603.

35. Liu, Y.; Wang, X.; Yang, F.; Yang, X. Excellent antimicrobial properties of mesoporous anatase TiO2 and Ag/TiO2 composite films. *Microporous Mesoporous Mater.* **2008**, *114*, 431–439.

36. Swetha, S.; Santhosh, S.M.; Geetha Balakrishna, R. Synthesis and comparative study of nano-TiO2 over degussa P-25 in disinfection of water. *Photochem. Photobiol.* **2010**, *86*, 628–632.

37. Dodd, N.J.F.; Jha, A.N. Photoexcitation of aqueous suspensions of titanium dioxide nanoparticles: An electron spin resonance spin trapping study of potentially oxidative reactions. *Photochem. Photobiol.* **2011**, *87*, 632–640.

38. Gabriel, M.M.; Mayo, M.S.; May, L.L.; Simmons, R.B.; Ahearn, D.G. *In vitro* evaluation of the efficacy of a silver-coated catheter. *Curr. Microbiol.* **1996**, *33*, 1–5.

39. Van Hullebusch, E.D.; Zandvoort, M.H.; Lens, P.N.L. Metal immobilisation by biofilms: Mechanisms and analytical tools. *Rev. Environ. Sci. Biotechnol.* **2003**, *2*, 9–33.

40. Slawson, R.M.; Vandyke, M.I.; Lee, H.; Trevors, J.T. Germanium and silver resistance, accumulation, and toxicity in microorganisms. *Plasmid* **1992**, *27*, 72–79.

41. Akhavan, O.; Ghaderi, E. Self-accumulated Ag nanoparticles on mesoporous TiO2 thin film with high bactericidal activities. *Surf. Coat. Technol.* **2010**, *204*, 3676–3683.

42. Kubacka, A.; Ferrer, M.; Cerrada, M.L.; Serrano, C.; Sanchez-Chaves, M.; Fernandez-Garci, M.; de Andres, A.; Rioboo, R.J.J.; Fernandez-Martin, F.; Fernandez-Garcia, M. Boosting TiO2-anatase antimicrobial activity: Polymer-oxide thin films. *Appl. Catal. B Environ.* **2009**, *89*, 441–447.

43. Jansen, B.; Kohnen, W. Prevention of biofilm formation by polymer modification. *J. Ind. Microbiol.* **1995**, *15*, 391–396.

44. Jiang, H.; Manolache, S.; Wong, A.C.L.; Denes, F.S. Plasma-enhanced deposition of silver nanoparticles onto polymer and metal surfaces for the generation of antimicrobial characteristics. *J. Appl. Polym. Sci.* **2004**, *93*, 1411–1422.

45. Pant, R.R.; Buckley, J.L.; Fulmer, P.A.; Wynne, J.H.; McCluskey, D.M.; Phillips, J.P. Hybrid siloxane epoxy coatings containing quaternary ammonium moieties. *J. Appl. Polym. Sci.* **2008**, *110*, 3080–3086.

46. Park, E.-S.; Lee, H.-J.; Park, H.Y.; Kim, M.-N.; Chung, K.-H.; Yoon, J.-S. Antifungal effect of carbendazim supported on poly(ethylene-*co*-vinyl alcohol) and epoxy resin. *J. Appl. Polym. Sci.* **2001**, *80*, 728–736.

47. Necula, B.S.; Fratila-Apachitei, L.E.; Zaat, S.A.J.; Apachitei, I.; Duszczyk, J. *In vitro* antibacterial activity of porous TiO2-Ag composite layers against methicillin-resistant Staphylococcus aureus. *Acta. Biomater.* **2009**, *2*, 3570–3580.

48. Jamuna-Thevi, K.; Bakar, S.A.; Ibrahim, S.; Shahab, N.; Toff, M.R.M. Quantification of silver ion release, *in vitro* cytotoxicity and antibacterialproperties of nanostuctured Ag doped TiO_2 coatings on stainless steel deposited by RF magnetron sputtering. *Vacuum* **2011**, *86*, 235–241.

49. Balamurugan, A.; Balossier, G.; Laurent-Maquin, D.; Pina, S.; Rebelo, A.H.S.; Faure, J.; Ferreira, J.M.F. An *in vitro* biological and anti-bacterial study on a sol-gel derived silver-incorporated bioglass system. *Dent. Mater.* **2008**, *24*, 1343–1351.

50. Sun, S.-Q.; Sun, B.; Zhang, W.; Wang, D. Preparation and antibacterial activity of $Ag-TiO_2$ composite film by liquid phase deposition (LPD) method. *Bull. Mater. Sci.* **2008**, *31*, 61–66.

51. Akhavan, O. Lasting antibacterial activities of $Ag–TiO_2/Ag/a-TiO_2$ nanocomposite thin film photocatalysts under solar light irradiation. *J. Colloid. Interface Sci.* **2009**, *336*, 117–124.

52. Kubacka, A.; Cerrada, M.L.; Serrano, C.; Fernández-García, M.; Ferrer, M.; Fernández-Garcia, M. Plasmonic nanoparticle/polymer nanocomposites with enhanced photocatalytic antimicrobial properties. *J. Phys. Chem. C* **2009**, *113*, 9182–9190.

53. Jung, W.K.; Koo, H.C.; Kim, K.W.; Shin, S.; Kim, S.H.; Park, Y.H. Antibacterial activity and mechanism of action of the silver ion in *Staphylococcus aureus* and *Escherichia coli*. *Appl. Environ. Microbiol.* **2008**, *74*, 2171–2718.

Antibacterial Activity of TiO₂ Photocatalyst Alone or in Coatings on *E. coli*: The Influence of Methodological Aspects

Thomas Verdier, Marie Coutand, Alexandra Bertron and Christine Roques

Abstract: In damp environments, indoor building materials are among the main proliferation substrates for microorganisms. Photocatalytic coatings, including nanoparticles of TiO₂, could be a way to prevent microbial proliferation or, at least, to significantly reduce the amount of microorganisms that grow on indoor building materials. Previous works involving TiO₂ have already shown the inactivation of bacteria by the photocatalysis process. This paper studies the inactivation of *Escherichia coli* bacteria by photocatalysis involving TiO₂ nanoparticles alone or in transparent coatings (varnishes) and investigates different parameters that significantly influence the antibacterial activity. The antibacterial activity of TiO₂ was evaluated through two types of experiments under UV irradiation: (I) in slurry with physiological water (stirred suspension); and (II) in a drop deposited on a glass plate. The results confirmed the difference in antibacterial activity between simple drop-deposited inoculum and inoculum spread under a plastic film, which increased the probability of contact between TiO₂ and bacteria (forced contact). In addition, the major effect of the nature of the suspension on the photocatalytic disinfection ability was highlighted. Experiments were also carried out at the surface of transparent coatings formulated using nanoparticles of TiO₂. The results showed significant antibacterial activities after 2 h and 4 h and suggested that improving the formulation would increase efficiency.

Reprinted from *Coatings*. Cite as: Verdier, T.; Coutand, M.; Bertron, A.; Roques, C. Antibacterial Activity of TiO₂ Photocatalyst Alone or in Coatings on *E. coli*: The Influence of Methodological Aspects. *Coatings* **2014**, *4*, 670-686.

1. Introduction

Indoor air pollution is a serious public health concern and a major cause of morbidity and mortality worldwide. In Europe, the total disease burden due to indoor air is about two million DALY (disability-adjusted life year) a year [1]. In 2006, the World Health Organization (Regional Office for Europe) started to draw up guidelines for indoor air quality [2] and addressed the three causes of indoor pollution that were most relevant for public health [3]:

- Biological indoor air pollutants (damp and mold) [4];
- Chemical indoor air pollutants (selected products) [5];
- Pollutants from indoor combustion of fuels (in progress).

The presence of microbial populations in damp indoor environments is one of the main causes of the degradation of indoor air quality and contributes to Sick Building Syndrome [6,7]. In Northern Europe and North America, the prevalence of mold contamination in buildings is estimated at between 20% and 40% [8]. Among the hundreds of microbial species that can be found in indoor environments [9–11], some are listed as potentially pathogenic species by the French High Council for Public Health and the France Environment Health Association [8,12,13]. Various studies have reported associations of mold growth with respiratory diseases in buildings, especially damp and water-damaged buildings [14]. Microorganisms may produce contaminants, *i.e.*, aerial particles, such as spores, allergens, toxins and other metabolites, that can be serious health hazards to occupants [15–23]. Frequent exposure to these contaminants can lead to various health troubles, including irritations and toxic effects, superficial and systemic infections, allergies and other respiratory or skin diseases [13,23–26]. Sick Building Syndrome has extensive economic and social impact [27–29]. A number of researchers have already pointed out that indoor building materials can become major sites of microbial growth when promoting conditions, such as high humidity and nutrient content, are present [30]. These conditions are easily satisfied in water-damaged buildings, damp buildings and badly-insulated buildings. Results from earlier studies have revealed that various microorganisms, including potentially pathogenic species, are detected on building materials [30].

A substantial amount of literature has been published on the effect of photocatalytic TiO_2 nanoparticles on microorganisms [31–34]. These studies show that the photocatalytic process in water is effective against a wide range of organisms, such as algae, viruses, fungi and bacteria. It should be noted that the different tests were carried out in aqueous slurry or with aqueous inoculum (sprayed or dropped), emphasizing the major role of water in the microorganism photo-killing process. In addition, TiO_2 nanoparticles can be used as (I) powder, usually dispersed in aqueous slurry or (II) film/coating applied to various substrates. Several works have highlighted very high bactericidal efficiency on different microorganisms: around 3 log after 30 min [35] and 6 log after 90 min [36] on *E. coli*, approximately 8 log after 90 min on mutans streptococci [37], *etc.* However, studies reporting such efficiencies used relatively strong light intensity, close to 10 W/m^2, and sometimes even beyond intensities in everyday use, up to 500 W/m^2, with photon wavelengths usually between 300 and 400 nm [38–40]. To our knowledge, no study reports such inactivation values with weaker light intensity, closer to a passive photocatalytic device. The efficiency of photocatalytic disinfection is attributed to the oxidative damage mainly induced by reactive oxygen species (ROS), such as $O_2^{\bullet-}$, H_2O_2 and $HO^{\bullet}$. These reactive oxygen species are produced by redox reactions between adsorbed species (such as water and oxygen) and electrons and holes photo-generated by the illumination of TiO_2. On the basis of studies on *Escherichia coli*, OH radicals were assumed to be the major cause of the bactericidal effect [41,42], although direct oxidation by "holes" (h+) from the valence band on the TiO_2 surface is also highlighted in some works [43,44]. Regarding the process of degradation, the authors agree that the outer membrane, if present (Gram-negative bacteria), is the first barrier and, once it is damaged, the cytoplasmic membrane is attacked. The loss of cytoplasmic membrane integrity, which is involved in the process of cellular respiration, leads to the death of the cell.

This work is a preliminary study on transparent coatings formulated using TiO_2 nanoparticles to fight against microbial proliferation in indoor conditions. As such, the first step of our work was to explore the different parameters influencing the efficiency of TiO_2 nanoparticles when used alone for disinfection, *i.e.*, before being included in coatings. The aim of the paper was to emphasize the different factors determining disinfection efficiency and to show that the various performances reported in the literature should be correlated with experimental parameters. Passive devices in the form of semi-transparent photocatalytic coatings, easy to apply to the building material surfaces, are also considered.

Our previous investigations have already shown the efficiency of semi-transparent coatings on the abatement of NO_x and VOC in air under various environmental conditions (Relative Humidity—RH, concentration of polluting gas, *etc.*) [45,46]. Such coatings consisted of ultra-light varnishes formulated using nanoparticles of TiO_2, acrylic resin and silicates as the inorganic binder. The results obtained in air purification point out the interest of testing these transparent coatings for the photocatalytic disinfection of microorganisms. However, the coatings were found to be inefficient against green algae colonization in accelerated tests [47]. Regarding TiO_2 nanoparticles alone, very good antibacterial performance is sometimes reported for photocatalytic TiO_2, but may be related to very specific experimental conditions that are not representative of the natural conditions to be considered for passive devices. Three sets of experiments were carried out to highlight different factors determining the extent to which *Escherichia coli*, a Gram-negative bacterium, was inactivated by TiO_2 photocatalysis: (1) the activity of TiO_2 in the dark allowed the photocatalytic effect to be dissociated from the physical effect; (2) the deposited drop experiment was carried out to evaluate the influence of forced conditions between bacteria and particles; and (3) the stirring experiment, which was easier to carry out for the kinetics evaluation, enabled the effect of the suspension to be estimated.

We also highlight some of the issues to be faced in the formulation of such a product, for example the inclusion of nanoparticles within a binder matrix (acrylic resin here), which can act as a mask against UV absorption and/or can react with photogenerated radicals.

2. Materials and Methods

2.1. Cultivation of Bacteria

Escherichia coli CIP 53126 was obtained from Institut Pasteur Collection, Paris, France. The strain was preserved at -80 °C in Eugon medium supplemented with 10% glycerol. Before each experiment, bacterial cells were pre-cultured on a nutrient agar slant. They were then transferred to a trypticase soy agar and incubated at a temperature of 36 °C $\pm$ 1 °C for 16 to 24 h. In addition, one plastic loop of bacteria was transferred to a fresh trypticase soy agar and incubated at a temperature of 36 °C $\pm$ 1 °C for 16 to 20 h prior to the test. For testing, one plastic loop of bacteria was dispersed evenly in a small amount of 1/500 nutrient broth (NB) [48] or of sterile distilled water, depending on the test, and the bacterial cell content of the suspension for inoculation was adjusted to about 10^8 cells/mL with a spectrophotometer (640 nm). The cell suspension was then 10-fold steps diluted, and 1 mL of each dilution was incorporated in trypticase soy agar to determine the number of CFU/mL. The test suspensions were prepared by 10-fold dilutions.

2.2. Antibacterial Activity of TiO_2 in the Dark

TiO_2 nanoparticles (KRONOClean 7050) were suspended in 1/500 NB [48] at the concentration of 13.9 g/L. Eleven milliliters of the suspension were then deposited onto a sterile Petri dish, so that the total area of the inside part of the dish was covered. The Petri dishes were placed in a sterile flow hood for air drying until the water had totally evaporated. A film of TiO_2 was visible at the bottom. Then, 11 mL of the inoculum (between 8×10^4 and 2×10^5 cells/mL) were deposited on the TiO_2 film, and the Petri dishes were covered with a lid [48]. After a fixed time (0 and 24 h), the lid was removed, the bottoms of the Petri dishes were gently scraped with a plastic loop in order to remove any adhered cells and 1 mL of the suspension was collected and diluted in phosphate buffer. Control samples were studied in Petri dishes without TiO_2.

One-mL quantities of the appropriate dilutions were then dropped into distilled sterile water and filtered on cellulose ester filters ($\phi = 0.45$ µm) in order to separate bacterial cells from nanoparticles. The filters were then deposited on trypticase soy agar and incubated at a temperature of 36 °C ± 1 °C for 40 to 48 h. After incubation, the number of viable cells was estimated in CFU/mL.

2.3. Deposited-Drop Experiment

To avoid damage by UV irradiation alone [49], the maximum UV intensity was maintained at 2.5 W/m^2. Previous tests with higher UV intensity had shown total drying of the inoculum during the experiment and led to the inactivation of bacteria in control samples. The light intensity was measured on the samples using a UV-A radiometer (Gigahertz-Optik, GmbH Türkenfeld, Germany) in the 310–400 nm range.

Various configurations were studied: samples under UV irradiation (TiO_2-bearing samples and control specimen without TiO_2) and samples kept in the dark (TiO_2-bearing samples and control specimen without TiO_2). All tests were carried out in triplicate. The data shown are the average of triplicates, with the corresponding standard errors.

2.3.1. With TiO_2 Powder

The experiment was based on the standards JIS Z 2801 (Japanese Industrial Standard) and ISO 27447 [48,49]. TiO_2 nanoparticle powder (KRONOClean 7050–anatase) was suspended in 9 mL of 1/500 NB [48], and 1 mL of the bacterial suspension (Section 2.1) was added. Final concentrations were 1 g/L for TiO_2 and between 8×10^4 and 2×10^5 CFU/mL for bacteria. The bacterial suspension (Section 2.1) without TiO_2 was used as a control. Then, 0.4 mL of the inoculum were instilled onto a Pyrex Petri dish designed so that an external ring could receive 2 mL of a supersaturated saline solution (KNO_3) to maintain 90% RH and was covered with a Pyrex lid (Figure 1). The Petri dishes were placed in a sterile flow hood and illuminated with an 8-W black-light bulb. After a few minutes, the TiO_2 nanoparticles were observed to have sedimented at the bottom of the drop.

A Soybean Casein Lecithin Polysorbate 80 Medium, also known as SCDLP broth, was prepared in sterile distilled water as recommended in standard JIS Z 2801 [48], using casein peptone, soybean peptone, sodium chloride, disodium hydrogen phosphate, glucose, lecithin and Tween 80.

Figure 1. Schematic illustration of the deposited-drop experiments with TiO₂ powder and TiO₂ semi-transparent coatings (**a**) and stirring experiment (**b**).

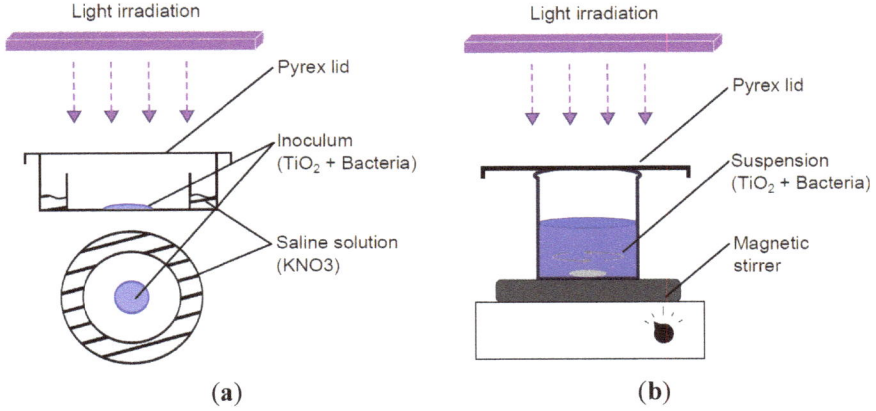

(**a**)   (**b**)

After different contact times (2 h, 4 h, 6 h), the suspension was washed out with the appropriate amount of SCDLP broth and with sterile glass beads (d = 4 mm). When necessary, the washed-out suspension was diluted X times in a phosphate buffer, so that it contained 30 to 300 cells per mL. For each sample, 1 mL of the appropriate dilution was dispensed into two sterilized Petri dishes with 15 to 20 mL of trypticase soy agar (TSA) and incubated at a temperature of 36 °C ± 1 °C for 40 to 48 h. After incubation, the number of viable cells was estimated in terms of CFU. The overall procedure was also systematically carried out immediately after inoculation (t = 0 h) to validate the test. The antibacterial activity was then calculated as the difference between the average logarithm of the number of viable bacteria on the control without TiO₂ and the average logarithm of the number of viable bacteria on the TiO₂ sample:

$$A = \log\left(N_{TiO_2}\right) - \log\left(N_{controrl}\right) = \log\left(\frac{N_{TiO_2}}{N_{controrl}}\right) \tag{1}$$

where, A: antibacterial activity; N_{TiO_2}: average number of CFU on TiO₂ sample at time t; $N_{control}$: average number of CFU on control sample at time t.

The test was then repeated with a transparent film (9–10 cm²) gently placed on the inoculum before irradiation in order to increase the probability of contact between bacteria cells and TiO₂ nanoparticles (forced contact).

2.3.2. With TiO₂ Semi-Transparent Coating

The deposited-drop experiment was repeated with semi-transparent coating formulated using TiO₂ nanoparticles as an antibacterial product: TiO₂ powder (KronoClean 7050, KRONOS/Société Industrielle du Titane, Paris, France) and TiO₂ dispersion (Kronos type 7454, trial product, KRONOS/Société Industrielle du Titane, Paris, France). The coating formulation included water and acrylic based on the work of Martinez *et al.* [46], as shown in Table 1. Sterilized cover-glasses (26 × 76 mm²) were covered with the coatings by instilling 1 mL, so that the total area of each glass was coated. The cover-glasses were then placed under a sterile flow hood for air drying. After drying, the semi-transparent coatings with TiO₂ powder (STC-SP (with silicates), STC-P) were gently sanded with fine sandpaper in order to prevent the possible inclusion of nanoparticles in the binder. The semi-transparent coatings with TiO₂ in aqueous suspension (STC-A) were pre-aged by irradiating them with UV light (2.5 W/m^2) for 80 h. The amount of TiO₂ was estimated at 2.5 mg/cm^2 for samples coated with TiO₂ powder (STC-SP, STC-P) and 0.63 mg/cm^2 for samples coated with TiO₂ aqueous suspension (STC-A). In order to evaluate the possible inclusion of nanoparticles in the binder of STC-A, samples were also prepared with water and TiO₂ aqueous suspension, without acrylic resin (STC-A2). For each test sample, corresponding controls were prepared in the same way with water and acrylic resin, but without TiO₂.

Table 1. Formulation of semi-transparent coatings. STC: semi-transparent coating.

STC-SP	STC-P	STC-A	STC-A2
Water	Water	Water	Water
Acrylic resin (7.5 wt%)	Acrylic resin (12 wt%)	Acrylic resin (2 wt%)	–
TiO₂ powder (KronoClean 7050)	TiO₂ powder (KronoClean 7050)	TiO₂ in aqueous suspension (Kronos trial product 7454)	TiO₂ in aqueous suspension (Kronos trial product 7454)
Silicates (12.5 wt%)	–	–	–

The inoculation suspension (of Section 2.1) was diluted to make the concentration of inoculum 8×10^4 to 2×10^5 CFU/mL. The coated cover-glasses were placed over the internal ring of the Pyrex Petri dishes shown in Figure 1. Relative humidity was maintained with 2 mL of the supersaturated saline solution (KNO₃) deposited in the external ring of each dish. Then, 0.4 mL of the inoculum were instilled on each coated cover-glass, and a transparent plastic film was applied, spreading the inoculum over a surface area of 10 cm^2. The Petri dishes were then covered with a Pyrex lid (Figure 1), placed in a sterile flow hood and illuminated with an 8-W black-light bulb.

After different contact times (2 h, 4 h, 6 h), the cover-glasses were recovered with sterile pliers and placed in plastic Petri dishes for wash-out. The wash-out of bacteria cells and the following procedures for CFU counting were repeated as in Section 2.3.1.

2.4. Stirring Experiment

For the stirring experiment; TiO₂ nanoparticles (KronoClean7050) were suspended in a sterile beaker in 27 mL of 1/500 NB or sterile distilled water; depending on the test; and 3 mL of the cell suspension were added to make the final test suspension. Final concentrations of the suspension were 1 g/L for TiO₂ and 7×10^4 to 1×10^5 CFU/mL for bacteria. A suspension of bacterial cells without TiO₂ was prepared as a control. The beakers (test sample and control) were placed in a sterile flow hood; covered with a Pyrex lid and illuminated with an 8-W black-light bulb at a light intensity of 5 W/m^2.

An aliquot of 1 mL was taken from each beaker every 30 min during 4 h and, when necessary, diluted in phosphate buffer before inclusion in trypticase soy agar as in the deposited-drop experiment. Controls were also carried out without TiO₂ and in the dark, with/without TiO₂. The data presented are the average of three experiments with the corresponding standard errors.

To assess the influence of the nature of the water during the experiment, two solutions were used: 1/500 nutrient broth and sterile distilled water. The two conductivities of the solutions were compared using a conductivity meter before the test. At room temperature (~21 °C), the conductivities were 4.437 mS/m for 1/500 NB and 1.1 mS/m for distilled water.

3. Results and Discussion

3.1. Effect of TiO₂ in the Dark

Figure 2 shows the bacterial concentration of *E. coli* cells after 0 and 24 h of contact with TiO₂ nanoparticles (11 mL inoculation of air-dried TiO₂ film in Petri dishes). From these data, an increase can be seen in the number of CFUs for control samples ($+2.25 \pm 0.06$ log) and a decrease in the number of CFUs for TiO₂ samples (-0.91 ± 0.14 log). The corresponding antibacterial activity, calculated from Equation (1), is 3.22 ± 0.14 log. It is therefore likely that the activity of TiO₂ on *E. coli* in the dark is correlated with a growth inhibitory effect as a major pathway and a bactericidal effect as a minor pathway. These results highlight the physical impact on *E. coli* cells induced by contact with TiO₂ nanoparticles, without regard to the photocatalytic process. This also agrees with earlier observations by Liu *et al.* [50] and Gogniat *et al.* [38], which showed a loss of bacterial culturability after contact with TiO₂ nanoparticles in the dark. A study by de Niederhäusen and Bondi [51] on the self-cleaning of Ag-TiO₂-coated ceramic tiles also showed significant antibacterial activity for 24 h in the dark.

Interestingly, we detected no difference in the CFU counts between bacterial suspensions with TiO₂ (1 g/L and 10 g/L) and a control bacterial suspension (without TiO₂) after direct plating of 2×1 mL on TSA and 48 hours' incubation (data not shown). It seems possible that the physical damage sustained is not sufficient to kill bacterial cells when they are growing in a nutrient-rich culture medium. Such a "neutralizing" effect of culture media is current with antiseptic and disinfectant molecules, which highlights the impact of test conditions on efficiency evaluation.

Figure 2. Bacterial concentration after 24 h with TiO₂ nanoparticles in the dark. Mean ± SE, $n = 3$.

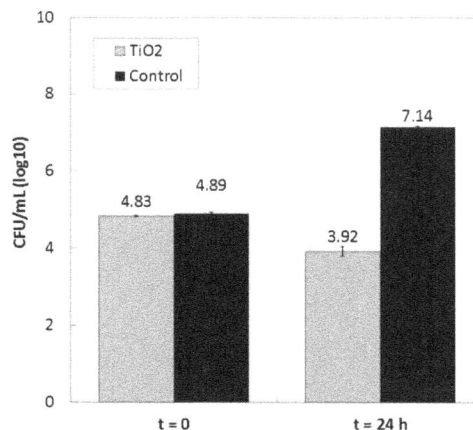

3.2. Free Surface Drop Deposit vs. Forced Contact between Bacteria and Nanoparticles

The antibacterial activities of TiO_2 nanoparticles on *E. coli* during the deposited-drop experiment are presented in Figure 3. No activity was detected for samples kept in the dark for 6 h under normal testing conditions (A). The bacterial reduction reached 0.92 ± 0.09 log after 6 h of irradiation (A).

Figure 3. Antibacterial activity of TiO_2 as a support under UV irradiation (≈ 2.5 W/m^2): (**A**) standard conditions; (**B**) after application of a transparent plastic film on the inoculum. Mean $\pm$ SE, $n = 3$.

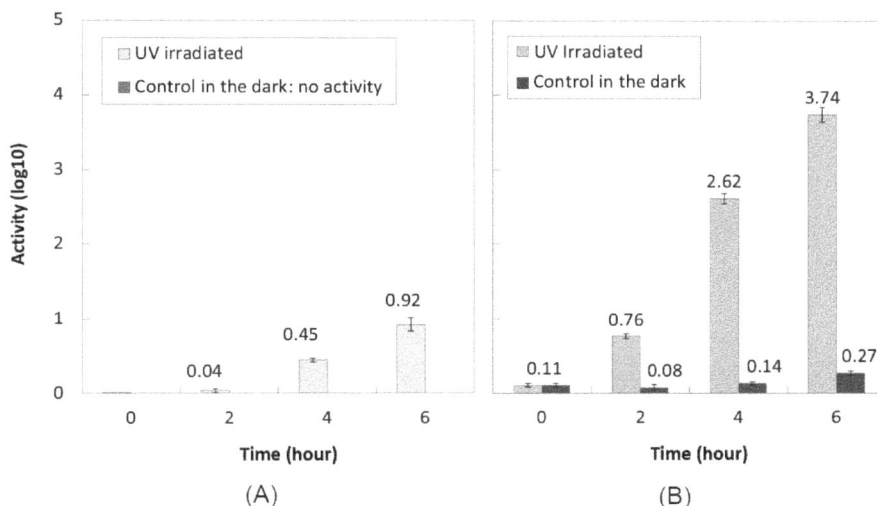

Since the bactericidal effect induced by photocatalysis of TiO_2 nanoparticles depends on many factors, such as the amount [35,37,52–55] and the crystalline nature [39,55–57] of TiO_2, the irradiation time and intensity [35,52,55,58] and the inoculum concentration [42,53,59], it is reasonable to assume that the dispersion of TiO_2 particles and bacteria (low probability of contact) and the very low light intensity used in our experiment (2.5 W/m^2 in order to avoid any UV-damage) could be major factors explaining the low activity observed after 6 h of irradiation. Various studies from the literature show intensities of over 10 W/m^2 and antibacterial activities on *E. coli* easily greater than 3 log after 90 min of irradiation [39,40,42,60].

Following the application of a transparent film (9 cm^2) onto the inoculum, a significant increase in the antibacterial activity was observed. As shown in Figure 3B, the activity was 3.74 ± 0.1 log after 6 h under UV irradiation and 0.27 ± 0.03 log after 6 h in the dark.

The present findings seem to support the idea that reducing the distance between bacterial cells and TiO_2 nanoparticles enhances the photocatalytic disinfection process. This also agrees with earlier research highlighting the importance of the contact between bacterial cells and the surface of TiO_2 [39–41,43,44,50,61–63]. In addition to the oxidative stress induced by reactive oxygen species (ROS) on bacterial cells, contact with the TiO_2 surface leads to the direct oxidation of cells by photogenerated holes, which also reduces the recombination of charges inside the photocatalyst [38,43,44,64]. Moreover, it has been suggested that direct contact and adsorption on

TiO₂ nanoparticles cause (I) a loss of membrane integrity [38,50,61] and possibly (II) a process of phagocytosis of the nanoparticles by the cells (the findings of Cai *et al.* [64] must be interpreted with caution in this paper, because they focused on HeLa cells and not bacterial cells.) [64], both leading to the reduction of the number of cultivable cells, if not to cell death. These results also agree with the findings of other studies that have highlighted the major role of surface radicals compared to free radicals in photocatalytic disinfection [40,41,62].

3.3. Influence of the Nature of the Solution for Suspension

The results obtained from the stirring experiment are presented in Figure 4. For TiO₂ samples, both inactivation curves consist of two steps: the first with a very low inactivation rate followed by the second with a higher inactivation rate. In addition, the second rate appears to be the same for both distilled water and 1/500 NB. It is likely that 1/500 NB acts as a retarding agent of the photocatalytic disinfection process.

Figure 4. Survival of *E. coli* cells *vs.* irradiation time at ~5 W/m² with the standard error of three experiments. Mean ± SE, $n = 3$.

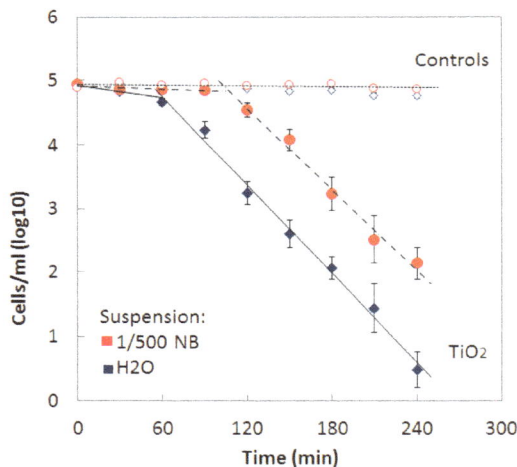

The present finding is in full agreement with the inhibitory effect of various ions and organic compounds on photocatalytic disinfection, which is widely reported in the literature [34,37,56,65,66]. The presence of ions and organic compounds can reduce the efficiency in different ways:

- Competition between ions, compounds and bacteria for the adsorption on the TiO₂ surface [37,38,56,65,67];
- The ROS mobilized by ions and compounds cannot oxidize bacterial cells [65];
- Aggregates of organic compounds could create a barrier filtering UV.

According to Dunlop *et al.* [56] and Sunada *et al.* [42], the low rate of inactivation in the first step may be due to the preliminary attack of the outer membrane of cells by ROS.

During this first step, the damage sustained by the outer membrane may be insufficient to kill bacteria: they can recover from the damage and re-grow once they are plated in agar media [42,56]. After some time, degradation of the outer membrane enables reactive species to penetrate, which induces damage, leading to the death of the bacterial cells (second, higher rates on the curves of Figure 4). This hypothesis has also been considered by other researchers [58,68]. Mitoraj *et al.* [68] explained this "incubation period" as the time for the concentration of photogenerated ROS to increase to a level that is harmful to bacteria.

Another possible explanation for the first step with the low inactivation rate is proposed by Gogniat *et al.* [38]. In their works, they observed the two-stage curve only in a sodium phosphate solution and not in a NaCl-KCl solution. They hypothesized that the change of adsorption properties of TiO_2 when illuminated led to a photo-desorption of ions previously adsorbed on its surface. Thus, the time taken for the photo-desorption process explains the low inactivation rate observed during the first minutes of the experiment [38].

Interestingly, the third step observed in earlier studies [58,68–70] and consisting of strong attenuation of the bacterial inactivation was not observed here. One of the hypotheses suggested is that photocatalytic inactivation is built up by bacterial growth after a certain period of time [69]. It can be supposed that bacterial growth in pure water is slowed down or stopped. Further investigations in 1/500 NB after longer times could show similar attenuation of the inactivation rate.

Some authors have compared efficiencies between scattered and fixed TiO_2 [71–73]. Pablos *et al.* [72] observed a higher inactivation rate at the beginning of the reaction with fixed TiO_2. They suggested that damage was uniformly distributed over the whole cell wall in slurries, whereas it was more concentrated on small areas with fixed TiO_2, requiring smaller amounts of radicals to achieve inactivation. However, they observed similar times for total inactivation of bacteria (*E. coli*) for both implementations (fixed and scattered). On the other hand, Gumy *et al.* [40] found higher inactivation efficiency with suspended TiO_2 than with TiO_2 coated on a fibrous web and suggested that particles dispersed in slurry would provide more surfaces for the adsorption of bacteria. In addition, inactivation of bacteria has been observed in the presence of TiO_2 nanoparticles in the dark, suggesting that phenomena other than photocatalytic processes can explain inactivation [50,61]. Although the complete process is not perfectly understood yet, the overall literature points to the importance of the contact between bacteria and TiO_2 for improving disinfection efficiency, suggesting both chemical and physical influences.

In their work, Gomes *et al.* [71] also reported a higher inactivation rate in slurry than with TiO_2 supported on Ahlstrom paper. They suggested that such results could be explained by competitive reactions of TiO_2 with the organic matter released by the paper during the experiment. Accordingly, the presence of ions and/or organic compounds in the slurry/inoculum considerably reduced the efficiency by reacting with ROS and being adsorbed on TiO_2 in place of bacterial cells [38,65]. These works also raised the problem of TiO_2 coatings in which the organic matter from the binder can monopolize photogenerated radicals and, thus, lead to a decrease in disinfection efficiency.

3.4. Semi-Transparent Coating

Figure 5 presents the experimental data on the antibacterial activity of semi-transparent coatings formulated with silicate and TiO_2 powder (STC-SP). Surprisingly, antibacterial activity was also observed on samples kept in the dark. A quick estimation of the pH of the inoculum with indicator paper showed a pH around 11–12, far too high for *E. coli* survival. It is then reasonable to assume that the antibacterial activity detected on STC-SP samples was not induced by the photocatalytic process, but by the silicates, making the inoculum strongly basic. Results from sample coatings without silicate (STC-P) showed no activity after eight hours' irradiation and no activity after 8 h in the dark. Possible explanations are that photogenerated radicals may have reacted with the binder instead of with bacteria or that the inclusion of nanoparticles within the binder may have prevented UV absorption and physical damage by contact. Moreover, the use of TiO_2 powder without a dispersing agent may lead to the formation of aggregates, reducing the surface available for reaction.

Figure 5. Antibacterial activity of SCT-SP coatings under UV irradiation ($\approx$2.5 W/m^2) and in the dark. Mean $\pm$ SE, $n = 3$.

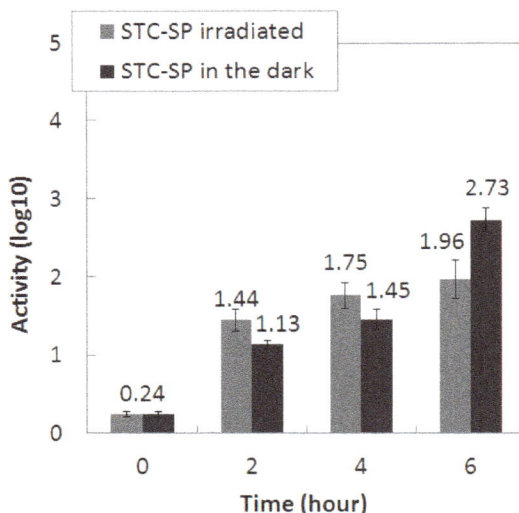

Figure 6 presents the antibacterial activity obtained with STC-A and STC-A2. The activity reaches 1.49 ± 0.47 log for STC-A and 1.54 ± 0.13 log for STC-A2 after four hours' irradiation. The observed increase of antibacterial activity, compared to SCT-P in which no activity was detected, could be attributed to the use of the TiO_2 dispersion. Nanoparticles, stabilized by the dispersing agent, may have provided more active sites for the photocatalytic process. Moreover, the smaller amount of acrylic resin within STC-A and STC-A2 (2%) may have reduced the inclusion of TiO_2 nanoparticles compared to SCT-P. The similar activities observed on STC-A (with acrylic resin) and on STC-A2 (without acrylic resin) seem to confirm this hypothesis. Further investigations on the formulation of these coatings along with observations of nanoparticle distribution in the binder will be helpful in the development of antibacterial products for building materials.

Figure 6. Antibacterial activity of STC-A and STC-A2 coatings under UV irradiation (≈ 2.5 W/m^2). Mean $\pm$ SE, $n = 3$.

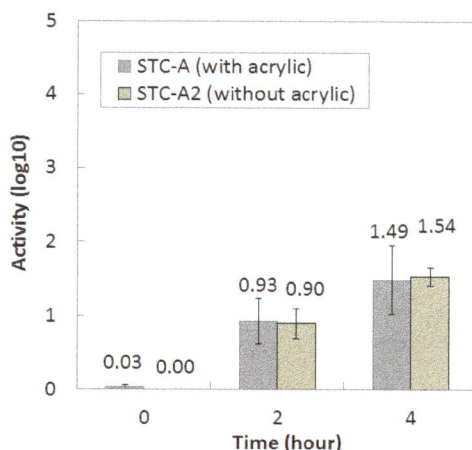

4. Conclusions

This paper has examined the effect of TiO$_2$ photocatalyst on *E. coli* in terms of antibacterial activity by carrying out two different tests (a drop deposited on a photocatalytic substrate and a stirring experiment in a TiO$_2$-bearing suspension).

Some general effects reported in the literature concerning the photocatalytic disinfection process have been observed.

- Prolonged contact (24 h) in the dark leads to significant antibacterial activities, potentially explained by a combination of the direct contact (I) bactericidal effect and the (II) growth inhibiting effect.
- Reducing the distance between nanoparticles and bacteria significantly increases the inactivation of *E. coli* by non-photocatalytic effects (direct contact) and the photocatalysis disinfection process.
- The presence of ions and organic compounds in the suspension during the test delays the inactivation.

In addition, the transparent coatings tested showed significant antibacterial activities under low UV irradiation. The results suggest that improving the formulation, *i.e.*, varying the proportions of the components, could increase the efficiency of coatings. In a broader framework, further experimental investigations will be conducted on the resistance of this coating to fungal proliferation and on the protection it affords against the formation of microbial biofilms.

Acknowledgments

The authors would like to thank Université Paul Sabatier Toulouse III for its financial support.

Author Contributions

Experimental measurements have been conducted by Thomas Verdier. Analysis and interpretation of the results as well as conclusions have been conducted by all the co-authors. The manuscript has been written by Thomas Verdier with the revision and approval by the others co-authors.

Conflicts of Interest

The authors declare no conflict of interest.

References

1. Déoux, S. *Les Enjeux Sanitaires de la Qualité de l'air Intérieur*; Cité des Sciences et de l'industrie: Paris, France, 2013. (In French)
2. Development of WHO Guidelines for Indoor Air Quality. Available online: http://www.euro.who.int/en/health-topics/environment-and-health/air-quality/publications/pre2009/development-of-who-guidelines-for-indoor-air-quality (accessed on 10 June 2014).
3. WHO Guidelines for Indoor Air Quality. Available online: http://www.euro.who.int/en/health-topics/environment-and-health/air-quality/policy/who-guidelines-for-indoor-air-quality (accessed on 10 June 2014).
4. World Health Organization. *WHO Guidelines for Indoor Air Quality: Dampness and Mould*; WHO: Geneva, Switzerland, 2009.
5. World Health Organization. *WHO Guidelines for Indoor Air Quality: Selected Pollutants*; WHO: Geneva, Switzerland, 2010.
6. Cooley, J.D.; Wong, W.C.; Jumper, C.A.; Straus, D.C. Correlation between the prevalence of certain fungi and sick building syndrome. *Occup. Environ. Med.* **1998**, *55*, 579–584.
7. Li, C.S.; Hsu, C.W.; Tai, M.L. Indoor pollution and sick building syndrome symptoms among workers in day-care centers. *Arch. Environ. Health* **1997**, *52*, 200–207.
8. Conseil Supérieur d'Hygiène Publique de France (CSHPF). *Contaminations Fongiques en Milieux Intérieurs. Diagnostic, Effet sur la Santé Respiratoire, Conduite à Tenir*; CSHPF: Paris, France, 2006. (In French)
9. Nolard, N.; Beguin, H. Moisissures. *Traité D'Allergologie Paris Médecine-Sci.*; Flammarion: Paris, France, 2003; pp. 441–461. (In French)
10. Gutarowska, B.; Żakowska, Z. Elaboration and application of mathematical model for estimation of mould contamination of some building materials based on ergosterol content determination. *Int. Biodeterior. Biodegrad.* **2002**, *49*, 299–305.
11. Minnesota Department of Health (MDH). *Recommended Best Practices for Mold Investigations in Minnesota Schools*; MDH, Environmental Health Division, Indoor Air Unit: Saint Paul, MN, USA, 2001.
12. Reboux, G.; Bellanger, A.-P.; Roussel, S.; Grenouillet, F.; Millon, L. Moisissures et Habitat: Risques Pour la Santé et Espèces Impliquées. *Rev. Fr. Allergol.* **2010**, *50*, 611–620. (In French)

13. ASEF Pollution de l'air intérieur de l'habitat. Available online: http://www.asef-asso.fr/attachments/1141_Guide_air%20int%C3%A9rieur.pdf (accessed on 30 October 2013). (In French)

14. Fung, F.; Hughson, W.G. Health Effects of Indoor Fungal Bioaerosol Exposure. *Appl. Occup. Environ. Hyg.* **2003**, *18*, 535–544.

15. Santucci, R.; Meunier, O.; Ott, M.; Herrmann, F.; Freyd, A.; de Blay, F. Contamination fongique des habitations: Bilan de 10 années d'analyses. *Rev. Fr. Allergol. Immunol. Clin.* **2007**, *47*, 402–408. (In French)

16. Nielsen, K.F.; Holm, G.; Uttrup, L.P.; Nielsen, P.A. Mould growth on building materials under low water activities. Influence of humidity and temperature on fungal growth and secondary metabolism. *Int. Biodeterior. Biodegrad.* **2004**, *54*, 325–336.

17. Spengler, J.D.; Chen, Q. Indoor Air Quality Factors in Designing a Healthy Building. *Annu. Rev. Energy Environ.* **2000**, *25*, 567–600.

18. Tuomi, T.; Reijula, K.; Johnsson, T.; Hemminki, K.; Hintikka, E.-L.; Lindroos, O.; Kalso, S.; Koukila-Kähkölä, P.; Mussalo-Rauhamaa, H.; Haahtela, T. Mycotoxins in Crude Building Materials from Water-Damaged Buildings. *Appl. Environ. Microbiol.* **2000**, *66*, 1899–1904.

19. Bellanger, A.-P.; Reboux, G.; Roussel, S.; Grenouillet, F.; Didier-Scherer, E.; Dalphin, J.-C.; Millon, L. Indoor fungal contamination of moisture-damaged and allergic patient housing analysed using real-time PCR. *Lett. Appl. Microbiol.* **2009**, *49*, 260–266.

20. Andersson, M.A.; Nikulin, M.; Köljalg, U.; Andersson, M.C.; Rainey, F.; Reijula, K.; Hintikka, E.L.; Salkinoja-Salonen, M. Bacteria, molds, and toxins in water-damaged building materials. *Appl. Environ. Microbiol.* **1997**, *63*, 387–393.

21. Dillon, H.K.; Miller, J.D.; Sorenson, W.G.; Douwes, J.; Jacobs, R.R. Review of methods applicable to the assessment of mold exposure to children. *Environ. Health Perspect.* **1999**, *107*, 473–480.

22. Torvinen, E.; Meklin, T.; Torkko, P.; Suomalainen, S.; Reiman, M.; Katila, M.-L.; Paulin, L.; Nevalainen, A. Mycobacteria and Fungi in Moisture-Damaged Building Materials. *Appl. Environ. Microbiol.* **2006**, *72*, 6822–6824.

23. *Health Implications of Fungi in Indoor Environments*; Samson, R.A., Flannigan, B., Flannigan, M.E., Verhoeff, A.P., Adan, O.C.G., Hoekstra, E.S.; Elsevier Science Ltd.: Kidlington, UK, 1994.

24. Flannigan, B.F.; Samson, R.A.; Miller, J.D. *Microorganisms in Home and Indoor Work Environments: Diversity, Health Impacts, Investigation and Control*; Taylor & Francis Group: Abingdon, UK, 2001.

25. Parat, S.; Perdrix, A.; Mann, S.; Cochet, C. A Study of the Relationship between Airborne Microbiological Concentrations and Symptoms in Office in Buildings. In Proceedings of the Healthy Building, Milan, Italy, 10–15 September 1995.

26. Williamson, I.J.; Martin, C.J.; McGill, G.; Monie, R.D.; Fennerty, A.G. Damp housing and asthma: A case-control study. *Thorax* **1997**, *52*, 229–234.

27. Mudarri, D.; Fisk, W.J. Public health and economic impact of dampness and mold. *Indoor Air* **2007**, *17*, 226–235.

28. Johanning, E. Mycotoxin and Indoor Health. In Proceedings of the Sixth VI International Conference on Mycotoxins in the Environment of People and Animals, Bydgoszcz, Poland, 25–27 September 2002.

29. Gutarowska, B.; Piotrowska, M. Methods of mycological analysis in buildings. *Build. Environ.* **2007**, *42*, 1843–1850.

30. Verdier, T.; Coutand, M.; Bertron, A.; Roques, C. A review of indoor microbial growth across building materials and sampling and analysis methods. *Build. Environ.* **2014**, *80*, 136–149.

31. Chong, M.N.; Jin, B.; Chow, C.W.K.; Saint, C. Recent developments in photocatalytic water treatment technology: A review. *Water Res.* **2010**, *44*, 2997–3027.

32. Gamage, J.; Zhang, Z.S. Applications of Photocatalytic Disinfection. *Int. J. Photoenergy* **2010**, *2010*, doi:10.1155/2010/764870.

33. Dalrymple, O.K.; Stefanakos, E.; Trotz, M.A.; Goswami, D.Y. A review of the mechanisms and modeling of photocatalytic disinfection. *Appl. Catal. B Environ.* **2010**, *98*, 27–38.

34. Foster, H.A.; Ditta, I.B.; Varghese, S.; Steele, A. Photocatalytic disinfection using titanium dioxide: Spectrum and mechanism of antimicrobial activity. *Appl. Microbiol. Biotechnol.* **2011**, *90*, 1847–1868.

35. Wei, C.; Lin, W.Y.; Zainal, Z.; Williams, N.E.; Hemminki, K.; Kruzic, A.P.; Smith, R.L.; Rajeshwar, K. Bactericidal Activity of TiO_2 Photocatalyst in Aqueous Media: Toward a Solar-Assisted Water Disinfection System. *Environ. Sci. Technol.* **1994**, *28*, 934–938.

36. Sunada, K.; Watanabe, T.; Hashimoto, K. Bactericidal Activity of Copper-Deposited TiO_2 Thin Film under Weak UV Light Illumination. *Environ. Sci. Technol.* **2003**, *37*, 4785–4789.

37. Saito, T.; Iwase, T.; Horie, J.; Morioka, T. Mode of photocatalytic bactericidal action of powdered semiconductor TiO_2 on mutans streptococci. *J. Photochem. Photobiol. B* **1992**, *14*, 369–379.

38. Gogniat, G.; Thyssen, M.; Denis, M.; Pulgarin, C.; Dukan, S. The bactericidal effect of TiO_2 photocatalysis involves adsorption onto catalyst and the loss of membrane integrity. *FEMS Microbiol. Lett.* **2006**, *258*, 18–24.

39. Gumy, D.; Morais, C.; Bowen, P.; Pulgarin, C.; Giraldo, S.; Hajdu, R.; Kiwi, J. Catalytic activity of commercial of TiO_2 powders for the abatement of the bacteria (*E. coli*) under solar simulated light: Influence of the isoelectric point. *Appl. Catal. B Environ.* **2006**, *63*, 76–84.

40. Gumy, D.; Rincon, A.G.; Hajdu, R.; Pulgarin, C. Solar photocatalysis for detoxification and disinfection of water: Different types of suspended and fixed TiO_2 catalysts study. *Sol. Energy* **2005**, *80*, 1376–1381.

41. Kikuchi, Y.; Sunada, K.; Iyoda, T.; Hashimoto, K.; Fujishima, A. Photocatalytic bactericidal effect of TiO_2 thin films: Dynamic view of the active oxygen species responsible for the effect. *J. Photochem. Photobiol. Chem.* **1997**, *106*, 51–56.

42. Sunada, K.; Watanabe, T.; Hashimoto, K. Studies on photokilling of bacteria on TiO_2 thin film. *J. Photochem. Photobiol. Chem.* **2003**, *156*, 227–233.

43. Nadtochenko, V.; Denisov, N.; Sarkisov, O.; Gumy, D.; Pulgarin, C.; Kiwi, J. Laser kinetic spectroscopy of the interfacial charge transfer between membrane cell walls of *E. coli* and TiO_2. *J. Photochem. Photobiol. Chem.* **2006**, *181*, 401–407.

44. Nadtochenko, V.A.; Sarkisov, O.M.; Nikandrov, V.V.; Chubukov, P.A.; Denisov, N.N. Inactivation of Pathogenic Microorganisms in the Photocatalytic Process on Nanosized TiO_2 Crystals. *Russ. J. Phys. Chem. B Focus Phys.* **2008**, *2*, 105–114.

45. Martinez, T. *Revêtements Photocatalytiques Pour les Matériaux de Construction: Formulation, Évaluation de L'efficacité de la Dépollution de l'air et Écotoxicité*; Génie Civil, Toulouse III–Paul Sabatier: Toulouse, France, 2012. (In French)

46. Martinez, T.; Bertron, A.; Ringot, E.; Escadeillas, G. Degradation of NO using photocatalytic coatings applied to different substrates. *Build. Environ.* **2011**, *46*, 1808–1816.

47. Martinez, T.; Bertron, A.; Escadeillas, G.; Ringot, E. Algal growth inhibition on cement mortar: Efficiency of water repellent and photocatalytic treatments under UV/VIS illumination. *Int. Biodeterior. Biodegrad.* **2014**, *89*, 115–125.

48. Japanese Industrial Standard. *Antibacterial Products—Test for Antibacterial Activity and Efficacy*; JIS Z 2801; Japanese Standards Association: Tokyo, Japan, 2010.

49. International Organization for Standardizaiton (ISO). *ISO 27447 Fine Ceramics (Advanced Ceramics, Advanced Technical Ceramics)—Test Method for Antibacterial Activity of Semiconducting Photocatalytic Materials*; ISO: Berlin, Germany, 2009.

50. Liu, L.; John, B.; Yeung, K.L.; Si, G. Non-UV based germicidal activity of metal-doped TiO_2 coating on solid surfaces. *J. Environ. Sci.* **2007**, *19*, 745–750.

51. De Niederhäusern, S.; Bondi, M.; Bondioli, F. Self-Cleaning and Antibacteric Ceramic Tile Surface. *Int. J. Appl. Ceram. Technol.* **2013**, *10*, 949–956.

52. Horie, Y.; David, D.A.; Taya, M.; Tone, S. Effects of Light Intensity and Titanium Dioxide Concentration on Photocatalytic Sterilization Rates of Microbial Cells. *Ind. Eng. Chem. Res.* **1996**, *35*, 3920–3926.

53. Maness, P.-C.; Smolinski, S.; Blake, D.M.; Huang, Z.; Wolfrum, E.J.; Jacoby, W.A. Bactericidal Activity of Photocatalytic TiO_2 Reaction: Toward an Understanding of Its Killing Mechanism. *Appl. Environ. Microbiol.* **1999**, *65*, 4094–4098.

54. Huang, Z.; Maness, P.-C.; Blake, D.M.; Wolfrum, E.J.; Smolinski, S.L.; Jacoby, W.A. Bactericidal mode of titanium dioxide photocatalysis. *J. Photochem. Photobiol. Chem.* **2000**, *130*, 163–170.

55. Rincón, A.G.; Pulgarin, C. Photocatalytical inactivation of *E. coli*: Effect of (continuous–intermittent) light intensity and of (suspended–fixed) TiO_2 concentration. *Appl. Catal. B Environ.* **2003**, *44*, 263–284.

56. Dunlop, P.S.M.; Byrne, J.A.; Manga, N.; Eggins, B.R. The photocatalytic removal of bacterial pollutants from drinking water. *J. Photochem. Photobiol. Chem.* **2001**, *148*, 355–363.

57. Caratto, V.; Aliakbarian, B.; Casazza, A.A.; Setti, L.; Bernini, C.; Perego, P.; Ferretti, M. Inactivation of Escherichia coli on anatase and rutile nanoparticles using UV and fluorescent light. *Mater. Res. Bull.* **2013**, *48*, 2095–2101.

58. Benabbou, A.K.; Derriche, Z.; Felix, C.; Lejeune, P.; Guillard, C. Photocatalytic inactivation of Escherischia coli: Effect of concentration of TiO_2 and microorganism, nature, and intensity of UV irradiation. *Appl. Catal. B Environ.* **2007**, *76*, 257–263.

59. Matsunaga, T.; Tomoda, R.; Nakajima, T.; Nakamura, N.; Komine, T. Continuous-sterilization system that uses photosemiconductor powders. *Appl. Environ. Microbiol.* **1988**, *54*, 1330–1333.

60. Kim, D.S.; Kwak, S.-Y. Photocatalytic Inactivation of *E. coli* with a Mesoporous TiO_2 Coated Film Using the Film Adhesion Method. *Environ. Sci. Technol.* **2009**, *43*, 148–151.

61. Caballero, L.; Whitehead, K.A.; Allen, N.S.; Verran, J. Inactivation of *Escherichia coli* on immobilized TiO_2 using fluorescent light. *J. Photochem. Photobiol. Chem.* **2009**, *202*, 92–98.

62. Pryor, W.A. Oxy-Radicals and Related Species: Their Formation, Lifetimes, and Reactions. *Annu. Rev. Physiol.* **1986**, *48*, 657–667.

63. Guillard, C.; Bui, T.-H.; Felix, C.; Moules, V.; Lina, B.; Lejeune, P. Microbiological disinfection of water and air by photocatalysis. *Comptes Rendus Chim.* **2008**, *11*, 107–113.

64. Cai, R.; Hashimoto, K.; Itoh, K.; Kubota, Y.; Fujishima, A. Photokilling of malignant cells with ultrafine TiO_2 powder. *Bull. Chem. Soc. Jpn.* **1991**, *64*, 1268–1273.

65. Rincón, A.-G.; Pulgarin, C. Effect of pH, inorganic ions, organic matter and H_2O_2 on *E. coli* K12 photocatalytic inactivation by TiO_2: Implications in solar water disinfection. *Appl. Catal. B Environ.* **2004**, *51*, 283–302.

66. Carp, O.; Huisman, C.L.; Reller, A. Photoinduced reactivity of titanium dioxide. *Prog. Solid State Chem.* **2004**, *32*, 33–177.

67. Okazaki, S.; Aoki, T.; Tani, K. The Adsorption of Basic α-Amino Acids in an Aqueous Solution by Titanium(IV) Oxide. *Bull. Chem. Soc. Jpn.* **1981**, *54*, 1595–1599.

68. Mitoraj, D.; Jańczyk, A.; Strus, M.; Kisch, H.; Stochel, G.; Heczko, P.B.; Macyk, W. Visible light inactivation of bacteria and fungi by modified titanium dioxide. *Photochem. Photobiol. Sci.* **2007**, *6*, 642–648.

69. Rincón, A.-G.; Pulgarin, C. Use of coaxial photocatalytic reactor (CAPHORE) in the TiO_2 photo-assisted treatment of mixed *E. coli* and Bacillus sp. and bacterial community present in wastewater. *Catal. Today* **2005**, *101*, 331–344.

70. Muranyi, P.; Schraml, C.; Wunderlich, J. Antimicrobial efficiency of titanium dioxide-coated surfaces. *J. Appl. Microbiol.* **2010**, *108*, 1966–1973.

71. Gomes, A.I.; Santos, J.C.; Vilar, V.J.P.; Boaventura, R.A.R. Inactivation of Bacteria *E. coli* and photodegradation of humic acids using natural sunlight. *Appl. Catal. B Environ.* **2009**, *88*, 283–291.

72. Pablos, C.; van Grieken, R.; Marugán, J.; Moreno, B. Photocatalytic inactivation of bacteria in a fixed-bed reactor: Mechanistic insights by epifluorescence microscopy. *Catal. Today* **2011**, *161*, 133–139.

73. Van Grieken, R.; Marugán, J.; Sordo, C.; Pablos, C. Comparison of the photocatalytic disinfection of *E. coli* suspensions in slurry, wall and fixed-bed reactors. *Catal. Today* **2009**, *144*, 48–54.

Tin Oxide-Silver Composite Nanomaterial Coating for UV Protection and Its Bactericidal Effect on *Escherichia coli* (*E. coli*)

Gil Nonato C. Santos, Eduardo B. Tibayan, Gwen B. Castillon, Elmer Estacio,

Takashi Furuya, Atsushi Iwamae, Kohji Yamamoto and Masahiko Tani

Abstract: SnO_2-Ag composite nanomaterials of mass ratio 1:4, 2:3, 3:2 and 4:1 were fabricated and tested for toxicity to *E. coli* using the pour-plate technique. The said nanomaterials were mixed with laminating fluid and then coated on glass slides. The intensity of UVA transmitted through the coated glass slides was measured. Results revealed that the 1:4 ratios of SnO_2-Ag composite nanomaterials have the optimum toxicity to *E. coli*. Furthermore, the glass slides coated with SnO_2 nanomaterial showed the lowest intensity of transmitted UVA.

Reprinted from *Coatings*. Cite as: Santos, G.N.C.; Tibayan, E.B.; Castillon, G.B.; Estacio, E.; Furuya, T.; Iwamae, A.; Yamamoto, K.; Tani, M. Tin Oxide-Silver Composite Nanomaterial Coating for UV Protection and Its Bactericidal Effect on *Escherichia coli* (*E. coli*). *Coatings* **2014**, *4*, 320-328.

1. Introduction

Glass materials in windows tend to act as good media for the growth and multiplication of microorganisms. The chemical constituents that are absorbed and deposited on glass windows provide nutrition to microorganisms; and thereby promote their growth [1]. The growth of microorganisms on the glass materials causes innumerable problems such as stains [2] and unacceptable odor [3] which is detrimental to human health [4]. It is therefore important to develop glass or glass coating materials with antimicrobial properties to protect the health of the populace.

On the other hand, it is known that ultraviolet radiation in the wavelength range of 280 to 320 nm, which is referred to as UVB, is harmful to the human skin. It causes sunburn, stains, or even skin cancer [5]. UVA, of wavelength range of 320–400 nm is harmful as well, causing skin aging and wrinkling (photoaging). It also causes photocarcinogenesis due to its ability to penetrate the dermal layers, unlike UVB which is absorbed by the epidermis [6]. The science and technology of protecting people against the harmful effects of UV radiation had received increasing interests most especially to countries with tropical climate wherein incidences of heat wave occurs [7]. Tinted glass windows are commonly used in buildings and vehicles to protect people from UV rays. However, the use of tints which are too dark is prohibited by law enforcement agencies since the use of such tints have been increasingly used in criminal activities [8–11]. It is therefore important to develop glass that is not tinted but is able to allow enough visible light to pass through. In this study, tin oxide-silver (SnO_2-Ag) composite nanomaterials were fabricated. Toxicity of the nanomaterials to *E. coli* bacteria was tested. The nanomaterials were mixed with laminating fluid. The mixtures were then coated on glass slides and allowed to dry. The ability of the coated glass slides to block UVA was monitored.

2. Experimental Section

2.1. Preparation of Nanoparticles

The raw materials used were Sigma Aldrich (St. Louis, MO, USA) Ag powder with 99.99% purity and <45 µm grain size and Merck SnO_2 powder of 99% purity with <5 µm grain size. The horizontal vapor phase crystal growth technique (HVPG) was used in the fabrication of SnO_2 nanomaterials, Ag nanomaterials, and SnO_2-Ag composite nanomaterials. In the said technique, 35 mg of SnO_2, Ag, and mixtures of SnO_2 with Ag powders at 1:4, 2:3, 3:2, and 4:1 mass ratio were placed into an ultrasonically clean closed-end quartz tubes with inner diameter of 8.5 mm, outer diameter of 11 mm, and length of 220 mm. The tube was evacuated using the Thermionics High Vacuum System with a vacuum pressure of 10^{-6} Torr which was sealed on the other end using by an Oxy-LPG gas mixture blow torch. The sealed quartz tubes were then placed in a Thermolyne horizontal tube furnace and were baked at a temperature of 800 °C with growth time of 6 h and ramp time of 80 min.

2.2. Characterization of Nanoparticles

The surface and elemental analysis of the SnO_2 nanomaterials, Ag nanomaterials, and SnO_2-Ag composite nanomaterials were carried out using JEOL SEM 5310 scanning electron microscope and Oxford EDX System, respectively. The pour-plate technique was then utilized to confirm the antimicrobial properties of the nanomaterials. In the said technique, *E. coli* bacterial solutions of 10^{-4} dilution factor were prepared through serial dilution from 0.5 Macfarland based standard solution. 2 mm of the said solutions were then poured into 6 opened quartz tubes. One quartz tube contained Ag nanomaterials, one quartz tube contained SnO_2 nanomaterials, while the other four tubes contained SnO_2-Ag nanomaterials of varying ratio of 1:4, 2:3, 3:2, and 4:1. Two additional quartz tubes were used, one containing Ag powder and another containing SnO_2 powder, in order to compare the effect on the bacterial activity of the bulk powders and nanomaterials. The quartz tubes containing the bacterial solutions were then shaken. One hundred microliters of the bacterial solution from each quartz tube was poured in separate sterile petri dishes. A 9 mL sterile and cold nutrient agar medium was then poured into each petri dish containing the bacterial solutions. The contents were thoroughly mixed and allowed to solidify. The dishes were then incubated at 35 °C for 24 h before comparing the colonies grown on each petri dish.

2.3. Characterization of Coatings

In performing the UVA test, 1 mg of Ag nanomaterials, SnO_2 nanomaterials, SnO_2-Ag composite nanomaterials, Ag powder, and SnO_2 powder were each mixed with 3mL of laminating fluid. Clean glass substrates were then coated with the nanomaterials-laminating fluid mixtures via the drip method. The UV optical properties of the coated glass slides were examined using the PASCO UVA light sensor in conjunction with an OMNI PAR 38 Flood 120 W lamp light source.

3. Results and Discussion

3.1. Surface Morphology and Elemental Composition of the Nanomaterials

Figure 1 shows the SEM images of (a) silver, (b) tin oxide, and (c,d) silver tin oxide nanomaterials grown at 800 °C with growth time of 6 h. The micrographs revealed the presence of nanoparticles, wires, and cotton-like structures grown in random directions.

Table 1 exhibits the atomic and elemental composition of the SnO_2-Ag composite nanomaterials at 1:4 ratio, 2:3 ratio, 3:2 ratio and 4:1 ratio. The resulting elemental and atomic composition confirmed the presence of Ag, Sn, and O in the composite nanomaterials.

Figure 1. SEM images of nanomaterials (**a**) silver; (**b**) tin oxide; and (**c,d**) silver tin oxide.

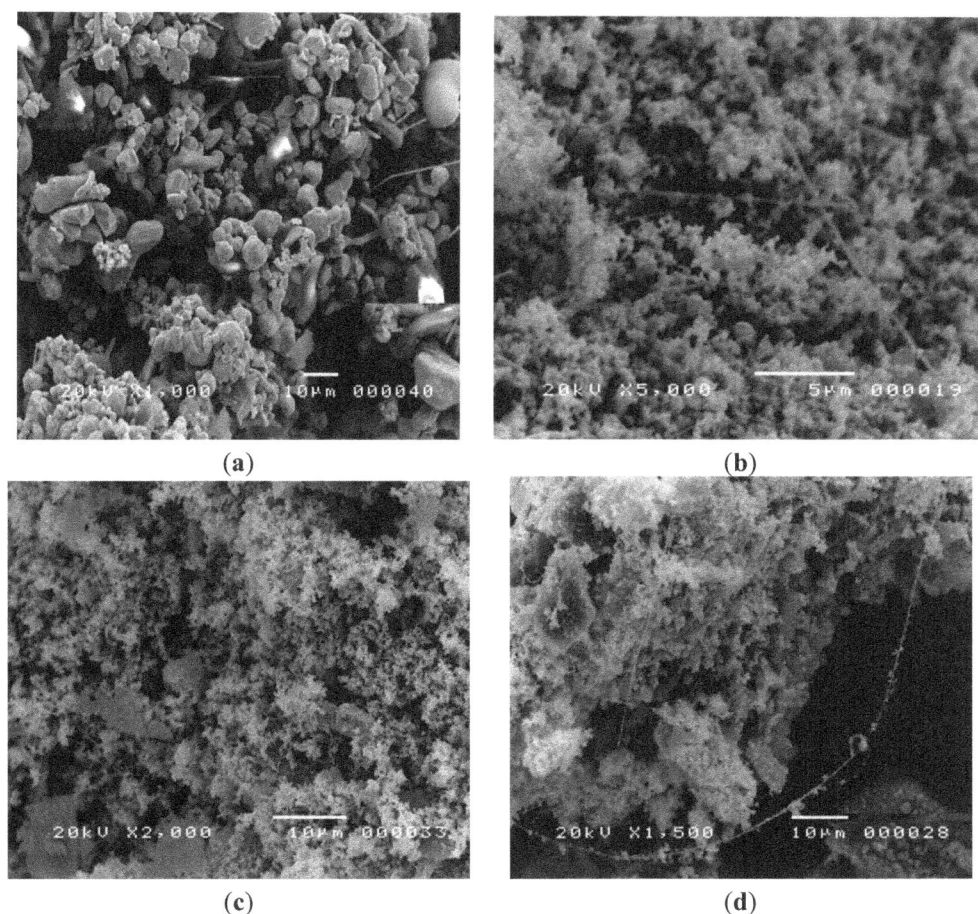

(**a**)

(**b**)

(**c**)

(**d**)

3.2. Antimicrobial Activity

The two controls used in the study are shown in Figure 2a,b. Figure 2a shows an agar plate without any bacterial colony. Instead of adding bacterial solution to the sterile petri dish, only distilled water was poured into it. On the other hand, Figure 2b shows an agar plate with several bacterial colonies seen as tiny white spots.

Table 1. Energy-dispersive X-ray spectroscopy (EDX) analysis of SnO_2-Ag nanomaterials at 1:4 mixture, 2:3 mixture, 3:2 mixture, and 4:1 mixture.

Ratio	Element	Elem%	Atom%
1:4	O	24.75	69.70
	Ag	45.84	19.14
	Sn	29.71	11.16
	Total	100.00	100.00
2:3	O	34.00	78.94
	Ag	20.27	4.37
	Sn	45.73	16.69
	Total	100.00	100.00
3:2	O	36.30	84.75
	Ag	10.38	3.18
	Sn	53.32	12.06
	Total	100.00	100.00
4:1	O	20.83	66.02
	Ag	3.59	1.69
	Sn	75.58	32.29
	Total	100.00	100.00

It can be seen in Figure 3 that there are fewer CFUs on the agar plate with Ag nanomaterials compared to the agar plate with the bulk Ag powder. Likewise, there are fewer CFUs on the agar plate with the SnO_2 nanomaterials than on the agar plate with bulk SnO_2 powder. This is consistent with the finding of Espulgar and Santos [12] that the antimicrobial property of bulk material was not only carried over but was enhanced by its nanomaterial counterpart.

Figure 4 shows that there are fewer CFUs for the agar plate with the 1:4 ratio of SnO_2-Ag composite nanomaterials as compared to the 2:3, 3:2, and 4:1 ratio of SnO_2-Ag composite nanomaterials.

Figure 2. Agar plates used as control for comparison: (**a**) Agar plate without bacteria; (**b**) Agar plate with bacteria.

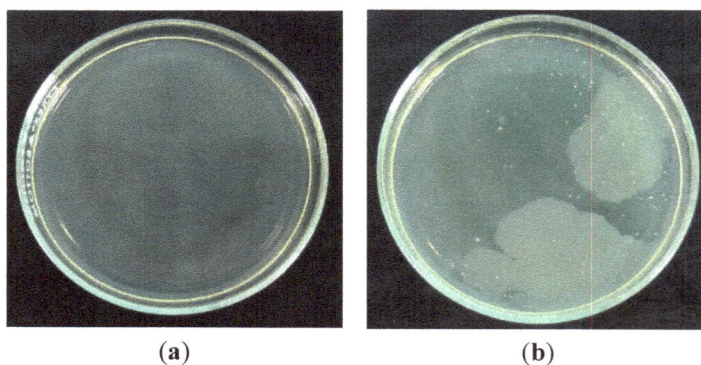

(**a**) (**b**)

Figure 3. Comparison on colony forming units (CFU) between (**a,c**) powder and (**b,d**) nanomaterial on (**a,b**) Ag and (**c,d**) SnO$_2$.

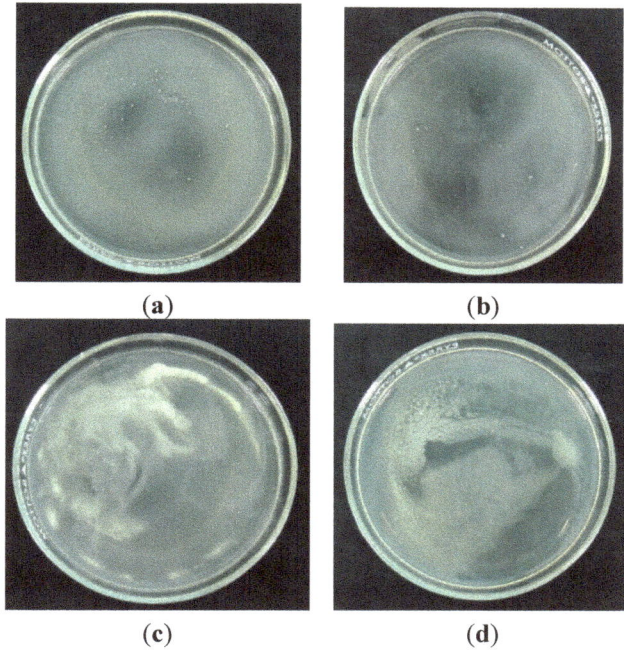

(a) (b)

(c) (d)

Figure 4. Agar plates containing mixtures of bacterial solution and nanomaterials of different ratio: (**a**) 1:4; (**b**) 2:3; (**c**) 3:2; (**d**) 4:1.

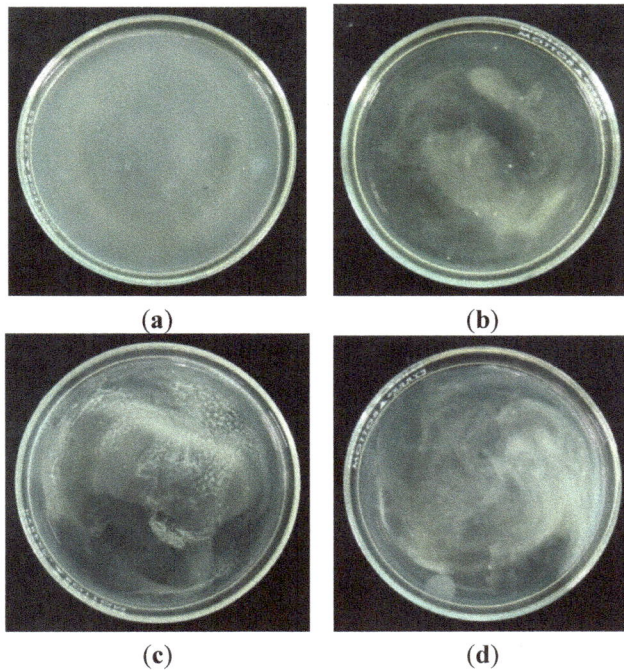

(a) (b)

(c) (d)

Table 2 summarizes the result of the antibacterial test where the number of CFU after 24 h of incubation was shown. It can be seen that Ag is more toxic to *E. coli* than SnO$_2$ and that the nanomaterials are more toxic than their bulk form. Also, it can be observed that as the percentage of Silver increases over tin oxide, the CFU number decreases. This is consistent with the observation that Ag is more toxic to *E. coli* than SnO$_2$. Moreover, results reveal that the 1:4 ratio of tin oxide and silver composite nanomaterials exhibits the greatest antimicrobial effect among the other ratios and material composition. Such a finding is consistent with previous reports [13,14] that the combination of Ag and a metal oxide may lead to an increase in bactericidal effect.

Table 2. Colony forming units (CFU) *vs.* material composition.

Material		CFU
Ag powder		59
Ag nanomaterial		18
SnO$_2$ powder		500
SnO$_2$ nanomaterial		35
SnO$_2$-Ag composite	1:4 ratio	9
	2:3 ratio	19
	3:2 ratio	20
	4:1 ratio	31
E. Coli		149

The mechanisms of the bactericidal effect of silver and silver nanoparticles (NPs) were discussed in different studies according to literature [15]. A study proposed that silver NPs can be attached to the surface of the cell membrane disturbing the permeability and respiration functions of the cell [4]. Smaller silver NPs having large surface area that are available for interaction would be more bactericidal than the larger silver NPs [16]. Moreover, it is possible that silver NPs will not only interact with the surface of membrane but can also penetrate inside the bacteria [17].

3.3. UVA Analysis of the Samples

Before the UVA transmission measurement was obtained, samples were first prepared by mixing 1 mg of Ag nanomaterials, SnO$_2$ nanomaterials, SnO$_2$-Ag composite nanomaterials, Ag powder, and SnO$_2$ powder with 3 mL of laminating fluid. Clean glass substrates were then coated with the nanomaterials-laminating fluid mixtures via the drip method. The coated glass slides were air dried and were later exposed to an OMNI PAR 38 Flood 120 W lamp light source. The relative intensity *vs.* time graph was obtained using a PASCO UVA light Sensor and was plotted using PASCO's Data Studio real time graph software. The graphs obtained are shown in Figure 5. As can be seen in the figure, the relative intensity was reduced when the glass slide was coated with the laminating fluid mixed with silver and tin oxide bulk powder. The basis for the decrease in intensity is that the material's surface was opaque and blocks UV light. On the other hand, the glass slide coated with the laminating fluid mixed with 4:1 ratio of tin oxide and silver nanomaterials had the least transmission for UVA and was transparent to visible light.

Figure 5. Graph of relative intensity of UVA *vs.* time graph of tin-oxide, silver, mass ratio of tin oxide silver nanocomposite material, glass slide, and without glass slide.

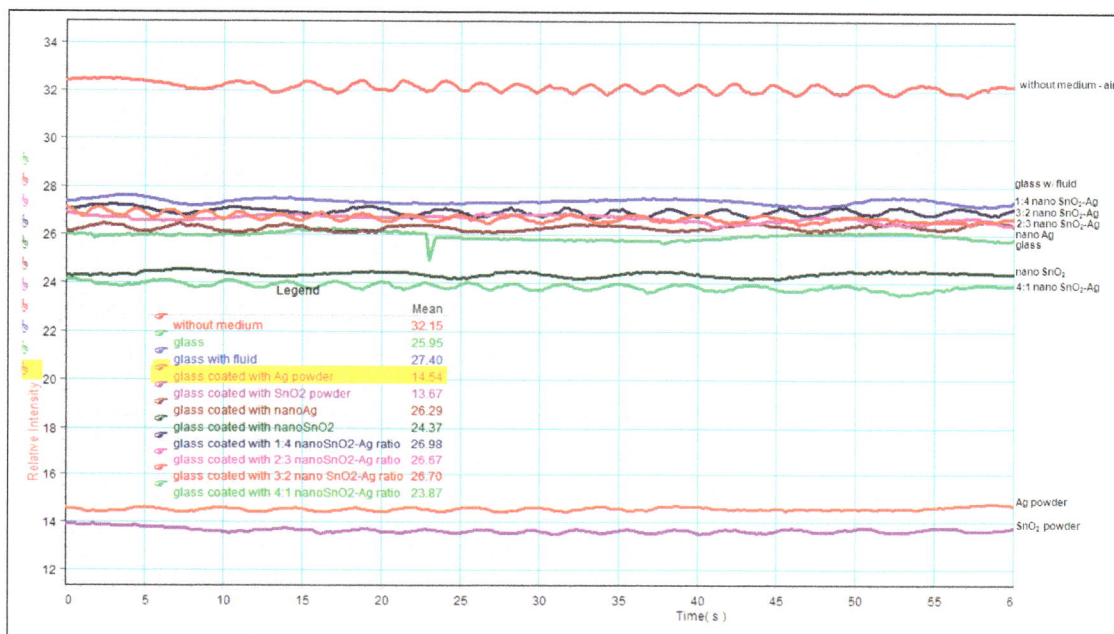

4. Conclusions

The HVPG technique was found to be effective in fabricating tin oxide and silver nanomaterials. The antimicrobial test shows that reducing the size of the sample to its nano form increases its toxicity to *E. coli* bacteria. Moreover, the glass slides coated with the laminating fluid mixed with tin oxide nanomaterial exhibits excellent UV blocking properties.

Acknowledgments

The researchers would like to thank DOST PCIEERD for the research grant for this project, M. Hangyo and K. Takano for the far infrared to mid-IR FT-IR measurements in their lab at the Institute of Laser Engineering, Osaka University. They also thank N. Miyoshi for his help for the near-IR FT-IR measurements in his lab at the Faculty of Medical Sciences, University of Fukui.

Author Contributions

Gil Nonato C. Santos is in charge of consolidation of the results and writing of the paper while Eduardo B. Tibayan and Gwen Castillon edited, synthesized and performed the antimicrobial tests. Elmer Estacio edited the paper regarding the optical characterization of the samples while Takashi Furuya, Atsushi Iwamae, and Kohji Yamamoto made the set up and performed the experiment of the samples using the FTIR and UV-VIS equipment. Masahiko Tani analyzed the UV blocking property of the samples.

48

Conflicts of Interest

The authors declare no conflict of interest.

References

1. Skarstad, K.; Steen, H.; Boye, E. Cell cycle parameters of slowly growing *Escherichia coli* studied by flow cytometry. *J. Bacteriol.* **1983**, *154*, 656–662.
2. Métris, A.; George, S.; Peck, M. Distribution of turbidity detection times produced by single cell-generated bacterial populations. *J. Microbiol. Methods* **2003**, *55*, 821–827.
3. Edwards, P.R.; Ewing, W.H. *Identification of Enterobacteriaceae*; Burgess Pub. Co.: Minneapolis, MN, USA, 1972.
4. Poindexter, J.S.; Leadbetter, E.R. *Methods and Special Applications in Bacterial Ecology*; Plenum Press: New York, NY, USA, 1986; p. 87.
5. Hockberger, P.E. A history of ultraviolet photobiology for humans, animals and microorganisms. *Photochem. Photobiol.* **2002**, *76*, 561–579.
6. Damiani, E.; Rosati, L.; Carloni, P.; Greci, L. Changes in ultraviolet absorbance and hence in protective efficacy against lipid peroxidation of organic sunscreens after UVA irradiation. *J. Photochem. Photobiol. B Biol.* **2006**, *82*, 204–213.
7. Matsumu, Y.; Ananthaswamy, H.N. Toxic effects of ultraviolet radiation on the skin. *Toxicol. Appl. Pharmacol.* **2004**, *195*, 298–308.
8. Pereira, V. Mangalore: Once Again Cops in Action to Impose Ban on Vehicles with Tinted Glass! Available Online: http://www.mangalorean.com/news.php?newstype=broadcast&broadcastid=450433 (accessed on 7 January 2014).
9. Impoundable Violations DOTC-LTO (MC 89–105). Metropolitan Manila Development Authority Website. Available online: http://www.mmda.gov.ph/faq.html#page-16 (accessed on 23 April 2014).
10. Harrow, C.M. Public Utility Vehicles Told to Remove Tint. Available Online: http://www.sunstar.com.ph/davao/local-news/public-utility-vehicles-told-remove-tints (accessed on 4 January 2011).
11. Agence France-Presse. Tinted Windows Ban Upsets Russian Drivers. Available Online: http://news.ph.msn.com/weird-news/tinted-windows-ban-upsets-russian-drivers-2 (accessed on 10 July 2012).
12. Espulgar, W.; Santos, G.N. Characterization of silver (Ag) nanomaterials synthesized by the HVPCG technique. *IJSER* **2011**, *2*, 126–129.
13. Yang, L.; Mao, J.; Zhang, X.; Xue, T.; Hou, T.; Wang, L.; Mingjing, T.U. Preparation and characteristics of Ag/nano-ZnO composite antimicrobial agent. *Nanosciences* **2006**, *1*, 44–48.
14. Hajipour, M.J.; Fromm, K.M.; Ashkarran, A.A.; de Aberasturi, D.J.; de Larramendi, I.R.; Rojo, T.; Serpooshan, V.; Parak, W.J.; Mahmoudi, M. Antibacterial properties of nanoparticles. *Trends Biotechnol.* **2012**, *30*, 499–511.

15. Morones, J.R.; Elechiguerra, J.L.; Camacho, A.; Holt, K.; Kouri, J.B.; Ramirez, J.T.; Yacaman, M.J. The bactericidal effect of silver nanoparticles. *Nanotechnology* **2005**, *16*, doi:10.1088/0957-4484/16/10/059.

16. Application of Nanotechnology in Lacquer Top Coat-Laboratory Studies. Available Online: http://www.primjetcolor.com.pl/badania_naukowe_ns_eng.html (accessed on 23 April 2014).

17. Lu, H.; Fei, B.; Xin, J.H.; Wang, R.; Li, L. Fabrication of UV-blocking nanohybrid coating via miniemulsion polymerization. *J. Colloid Interface Sci.* **2006**, *300*, 111–116.

Chapter 2:
Photocatalytic Removal of Dyes and Wettability

The Effect of Tween® Surfactants in Sol-Gel Processing for the Production of TiO₂ Thin Films

Ann-Louise Anderson and Russell Binions

Abstract: Titanium dioxide thin films were deposited using a Tween® surfactant modified non-aqueous sol-gel method onto fluorine doped tin oxide glass substrates. The surfactant concentration and type in the sols was varied as well as the number of deposited layers. The as deposited thin films were annealed at 500 °C for 15 min before characterisation and photocatalytic testing with resazurin intelligent ink. The films were characterised using scanning electron microscopy, atomic force microscopy, X-ray diffraction, Raman spectroscopy and UV-Vis spectroscopy. Photocatalytic activity of the films was evaluated using a resazurin dye-ink test and the hydrophilicity of the films was analysed by water-contact angles measurements. Characterisation and photocatalytic testing has shown that the addition of surfactant in varying types and concentrations had a significant effect on the resulting thin film microstructure, such as changing the average particle size from 130 to 25 nm, and increasing the average root mean square roughness from 11 to 350 nm. Such structural changes have resulted in an enhanced photocatalytic performance for the thin films, with an observed reduction in dye half-life from 16.5 to three minutes.

Reprinted from *Coatings*. Cite as: Anderson, A.-L.; Binions, R. The Effect of Tween® Surfactants in Sol-Gel Processing for the Production of TiO₂ Thin Films. *Coatings* **2014**, *4*, 796-809.

1. Introduction

Semiconductor materials are of long standing research interest due to their practical applications as photocatalysts for environmental remediation, with specific focus on titanium dioxide owing to its durability and high performance. Titanium dioxide has been used consistently as the photocatalyst of choice to address a variety of environmental problems, and has been shown to be particularly effective when utilised as a thin film coating, specifically in self-cleaning glass [1], antimicrobial applications [2], and for water-splitting to produce hydrogen [3]. A multitude of approaches are used to produce TiO₂ thin films, including sol-gel [4] and hydrothermal routes [5], as well as vapour deposition methods, such as chemical vapour deposition (CVD) [6], physical vapour deposition (PVD) [7] and more recently, aerosol-assisted CVD [8] and electric field assisted CVD [9].

Of all the deposition methods mentioned, sol-gel remains one of the most popular routes for producing TiO₂ thin films due to its low cost, experimental simplicity and easy scale-up ability. Sol-gel also enables direct control of particle homogeneity during the particle growth phase, and as a result it is a particularly popular strategy for simple modification of thin films for the properties listed above. There has been a great variety of research focused on process modification that encourages specific morphological control within the resulting thin films [10]. Typically, sol-gel methods have been modified with the addition of block co-polymer templating agents [11] or non-ionic surfactants, such as Triton X-100 [12]. Previously we have reported the use of Brij® surfactants in a non-aqueous sol-gel process to produce TiO₂ thin films with an increased average particle size and increased surface roughness, whereby such structural changes led to an increase in the photocatalytic activity of produced TiO_2 thin films [13]. The use of such non-ionic surfactants in sol-gel processing is a commonplace strategy for the direct control of particle size and shape during the growth phase for the enhancement of resulting properties [14,15]. Due to their amphiphilic nature, surfactants act as pore-directing agents that can enable the production of highly porous materials with specific pore size and structure. A wide variety of surfactants have been used in sol-gel processing for TiO_2 thin film production including Brij® surfactants [14,16], Triton™ X-100 [17], Pluronic triblock copolymers [14,15] and Tween® 20 [18], which has been used in this investigation in comparison with Tween® 40.

This paper focuses on the use of Tween® 20, 40 surfactants in a modified non-aqueous sol-gel method to investigate the effect of surfactant type and concentration on the subsequent microstructure and functional properties of TiO_2 thin films.

2. Experimental Details

2.1. Sol Synthesis

All chemicals were purchased from Aldrich and used without further purification. Two variations of Tween® non-ionic long chain surfactants were used as pore directing agents for sol-gel processing; Tween® 20 and Tween® 40. Each surfactant was added in varying concentrations as listed in Table 1. Preparation of the TiO_2 sol involved initial incorporation of the surfactant into the solvent, isopropanol (69 mL) with vigorous stirring until a homogeneous suspension was obtained. Acetylacetone (0.61 mL) was then added dropwise with continuous stirring, to act as a stabilising agent in order to control the speed of the hydrolysis and subsequent condensation reactions. The titania precursor, titanium tetraisopropoxide (TTIP, 6 mL) was then added dropwise to the solution with continuous stirring. The resulting transparent solution was stirred for a further 30 min before adding acetic acid (6.86 mL) dropwise. The solution was stirred for an additional 30 min, and then the sol was ready for use in dip-coating. The molar ratio of surfactant/TTIP/acetylacetone/isopropanol/acetic acid was R 0.2:0.06:9:0.04, where the surfactant concentration R was varied from 0 to 0.0006 mol·dm^{-3}. Sols were typically transparent, pale yellow in colour, homogeneous and stable for several weeks.

2.2. Sol-Gel Dip-Coating

A homemade dip-coating apparatus was used to prepare the TiO_2 thin films on glass substrates (F-SnO$_2$, 1.5 cm × 6 cm, Pilkington-NSG, St Helens, UK). The substrates were dip-coated into the sol with a controlled removal speed of 2.8 cm/min. The number of dips was varied to include samples with one to three layers by repeating the number of dips. Before each additional layer was added, the substrates were left under a fume hood for 30 min to allow the solvent to evaporate. After dip-coating, the samples were left overnight under a fume hood to evaporate any remaining solvent. Samples were annealed using a Carbolite Type 301 programmable furnace (Carbolite, Hope, UK) at 500 °C for 15 min with a ramp rate of 288.15 K·min^{-1}.

Table 1. Sol gel names with surfactants used and relevant concentrations with number of layer and annealing temperature and times.

Surfactant	Sol name	No. of layers	Concentration of surfactant (mol·dm⁻³)	Annealing temperature and time
Nil	A	1 2 3	0	500 °C/15 min
Tween® 40	B	1 2 3	0.0006	500 °C/15 min
	C	1 2 3	0.0003	500 °C/15 min
Tween® 20	D	1 2 3	0.0006	500 °C/15 min
	E	1 2 3	0.0003	500 °C/15 min

2.3. Materials Characterisation

The thin film samples produced were characterised and analysed using a range of instruments. SEM images were produced using an FEI Inspect F Scanning Electron Microscope (SEM) (FEI, Hillsboro, OR, USA) to investigate thin film microstructure, surface morphology, and film thickness from sample cross sections. ImageJ 1.48 [19] software was used to determine the average particle size from SEM images. Atomic Force Microscopy was used to determine the surface roughness of the thin films and analysis was completed using NT-MDT NTEGRA (Zelenograd, Moscow, Russia). Semi-contact mode imaging was performed under ambient conditions in air using silicon tips (Acta-20-Appnano ACT tapping mode with aluminium reflex coating, Nanoscience Instruments, Chicago, IL, USA) with Resonant Frequency of 300 kHz and Spring constant of 40 N·m⁻¹. Scan resolution of 256 samples per line. Images were processed and analysed by the offline software Nova 1.0.26.1443. X-ray Diffraction (Panalytical, Spectris, Egham, UK) was used to determine the crystalline phases of the TiO_2 thin films, using a Panalytical X'Pert Pro diffractometer in a glancing angle ($\alpha = 3°$) mode using a CuKα X-ray source (Kα1 = 0.1540598 nm; Kα2 = 0.15444260 nm. The diffraction patterns were collected over 10°–70° with a step size of 0.03° and a step time of 1.7 s·point⁻¹. Raman Spectroscopy was used for further determination of crystalline phases and impurities using a Renishaw (Wooton-under-Edge, UK) Raman system 1000 with helium neon laser of wavelength 514.5 nm. Contact angle measurements were used to determine the hydrophilicity of the thin film surfaces, by measuring the contact angle of deionised water before and after 30 min of UV irradiation with a 254 nm UV lamp (2 × 8 W-254 nm Tube, Power: 32 W), and a Goniometer Kruss DSA100 drop shape analyser (Kruss, Hamburg, Germany). Band gaps were determined using the Tauc method [20].

2.4. Resazurin Intelligent Ink for Photocatalytic Testing

Intelligent ink based on the dye Resazurin (Rz) was used to determine the photocatalytic activity of the TiO_2 thin films. This dye is commonly used as an indicator ink for the determination of photocatalytic activity due to its novel photo-reductive mechanism that can be monitored visually through a colour change from blue to pink upon degradation, and also quantitatively by following the degradation of the resazurin peak over time using UV-Vis spectroscopy (Perkin Elmer, Waltham, MA, USA) [21]. The ink was sprayed onto the TiO_2 surface and the degradation of the dye monitored before and after UV irradiation at 365 nm. The intelligent ink was made from deionized water (40 mL), hydroxyethyl cellulose polymer (3 g, 1.5 wt.% aqueous solution, Aldrich, Poole, UK), glycerol (0.3 g, Aldrich) and Resazurin (4 mg, Aldrich). The samples were sprayed with the ink solution using an aerosol spray gun and subsequently irradiated at a distance of 15 cm with a 365 nm UV lamp (2 × 8 W-365 nm Tube, Power: 32 W) at incremented times so that the photocatalytic degradation could be monitored closely. For spray coating the glass substrates were mounted on a sheet of paper and placed vertically on a wall in a fume hood as previously reported by Kafizas *et al.* [21], the thin film surface was coated in stages by spraying the substrate horizontally from left to right in short even steps, to ensure coverage of the entire substrate. This process was repeated several times until a pale blue coating was clearly visible. The resulting UV-Vis spectroscopy data was normalised utilising additional data collected using a blank glass substrate coated with the intelligent ink without a TiO_2 thin film, and a clean glass substrate without a TiO_2 thin film. The half-life could then be found from the normalised data to give the values quoted in Table 2.

3. Results and Discussion

The TiO_2 thin films produced by the surfactant-assisted sol-gel method described were transparent, covering the entire surface of the area of glass that was dipped and showed evidence of birefringence. All films produced exhibited good adherence to the substrate after annealing, and passed the scotch tape test. The average thickness of TiO_2 thin films produced varied within the range of 42–220 nm, whereby the thinnest film at 42 nm was a one layer sample produced from a sol-gel solution that did not contain any added surfactant, and the thickest film, 220 nm was a three layer sample produced from a sol-gel solution with 0.003 $mol \cdot dm^{-3}$ of Tween® 20 surfactant added. The thickness of films produced was found to vary depending on the surfactant and concentrations used during processing, as well as the number of layers. The thin film thicknesses are shown in Table 2, where it can be seen that on average the thickness of the film increases with consecutive number of dips, as expected as this increases the number of layers of TiO_2 on the surface. Single layer samples varied between 70–200 nm, two layer samples were within the thickness range of 110–210 nm, and three layer samples varied between 130–220 nm.

Table 2. Samples prepared via sol-gel using different types and concentrations of Tween® surfactant, and annealed at 500 °C for 15 min. A = no surfactant; B = Tween® 60 (6 × 10^{-4} mol·dm^{-3}); C = Tween® 60 (4 × 10^{-4} mol·dm^{-3}); D = Tween® 60 (2 × 10^{-4} mol·dm^{-3}); E = Tween® 40 (6 × 10^{-4} mol·dm^{-3}); F = Tween® 40 (4 × 10^{-4} mol·dm^{-3}); G = Tween® 20 (6 × 10^{-4} mol·dm^{-3}); H = Tween® 20 (4 × 10^{-4} mol·dm^{-3}). Numbers in sample name represent number of layers. Particle sizes marked with asterisks (*) denote agglomeration within the thin film. Contact angle measurements are given with standard deviation values.

Sample name	Contact angle (°) before UV	Contact angle (°) after UV	Average particle size with average deviation (nm)	Average root mean square roughness (nm)	Average thin film thickness (nm)	Photocatalytic half-life for Rz (min)	Indirect band gap (eV) ±0.05
A1	38.06 ± 9	14 ± 1	130 ± 30	9	42	11	2.9
A2	49.59 ± 5	6.94 ± 1	130 ± 28	11	78	16.5	3.1
A3	39.17 ± 2	7.42 ± 5	40 ± 32	17	91	9.5	3.2
B1	48.81 ± 2	4.69 ± 2	55 ± 33	318	180	–	3.2
B2	51.97 ± 1	8.67 ± 8	28 ± 21	196	170	5	3.2
B3	62.21 ± 1	7.78 ± 1	28* ± 21	209	210	–	3.0
C1	45.42 ±3	4.2 ± 2	39 ± 40	296	200	–	3.2
C2	36.66 ± 1	6.3 ± 4	29 ± 37	135	110	3	3.2
C3	61.72 ± 7	55.5 ± 1	46 ± 17	254	190	–	3.2
D1	37 ± 1	6.31 ± 2	40 ± 33	293	110	–	3.2
D2	36.61 ± 2	3.23 ± 1	25 ± 21	350	160	4	3.2
D3	47.3 ± 7	4.55 ± 2	40 ± 28	366	180	–	3.1
E1	33.71 ± 9	7.94 ± 1	41 ± 39	144	120	–	3.15
E2	36.82 ± 4	3.83 ± 1	35 ± 37	191	210	3	3.15
E3	34.35 ± 9	4.55 ± 2	26 ± 18	181	220	–	3.15

Furthermore, the addition of surfactant was found on average to produce thicker thin films. For example, the three layer sample without surfactant, A3 had a thickness of 130 nm, whereas three layer samples produced with surfactant ranged in thickness depending on the type and concentration used; 180–220 nm, as shown in Table 2. The addition of surfactant increases the viscosity of the original sol-gel solution, and further increasing the surfactant concentration causes additional viscosity within the sol-gel, so the resulting thin films are thicker [12].

The thin film microstructure can be analysed using the SEM images, whereby it was found that all variants explored; the addition of different surfactants, variation in the number of dips, and the annealing temperature, all influenced the morphology of the TiO2 thin films produced as shown in Figures 1–3. These variations within the sol-gel processing used were also found to significantly affect the photocatalytic and wetting properties of the thin films. Surfactant addition was found to alter the morphology of the TiO2 thin films produced, as seen in the SEM images in Figures 1–3. In comparison to the thin films produced without surfactant, as in Figure 1, those films produced with the addition of Tween® surfactants (Figures 2 and 3) show less agglomeration, greater particle definition and exhibited a wider particle size range within samples.

Figure 1. SEM Images of samples prepared via sol-gel. A = no surfactant; numbers represent number of layers. Samples were annealed at 500 °C/15 min.

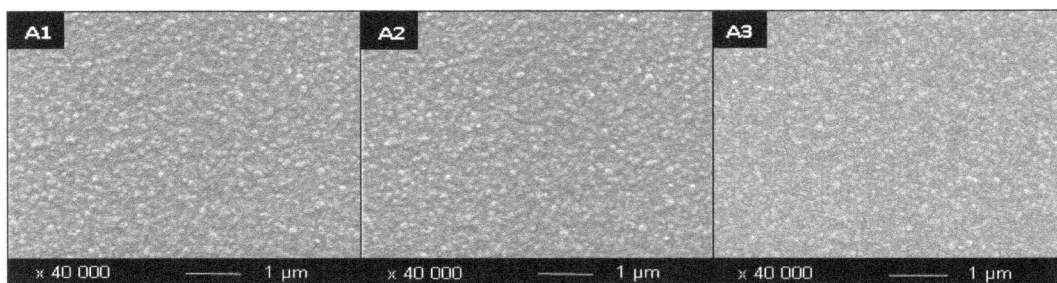

Figure 2. SEM Images of samples prepared by sol-gel with decreasing concentration of Tween® 40 surfactant. B = 0.0006 mol·dm^{-3}; C = 0.0003 mol·dm^{-3}; numbers represent number of layers. Samples were annealed at 500 °C/15 min.

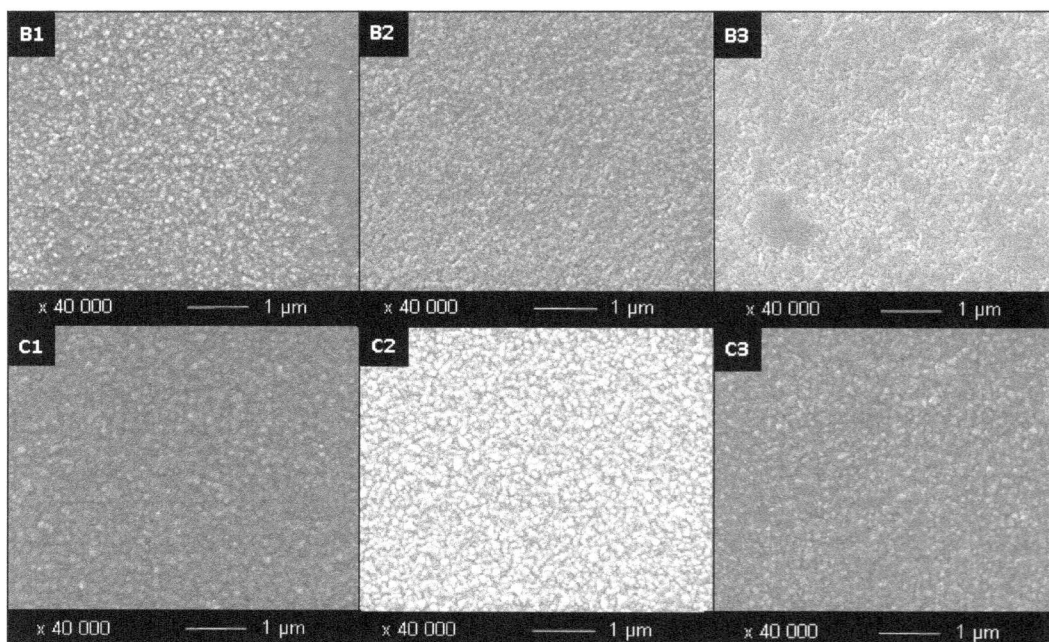

3.1. Surfactant Influence on the Morphology of Thin Films

The addition of Tween® surfactants was found to exhibit a range of effects on the morphology of the thin films produced, depending on the concentration and type of surfactant used. Surfactant addition was found to decrease average particle size, from 130 nm for two layer samples produced without surfactant, to as low as 25 nm for two layer samples produced with Tween® 20 surfactant at the higher concentration of 6×10^{-4} mol·dm^{-3}. The smaller particle sizes as listed in Table 2 are attributable to the role of the surfactant during the sol-gel growth phase, whereby the surfactant orients itself around growing titania particles restricting their growth to produce smaller particles. In addition, the samples produced with the addition of surfactant show greater particle definition and less agglomeration compared to the samples produced without surfactant (Figures 1–3) whereby the particles are also more angular due to the surfactant restricting their growth in a random way. This reduced particle size and enhanced particle definition increases the resulting surface area to volume ratio within the TiO_2 thin film sample. This leads to improved functional properties, such as improved photocatalytic activity, which has been shown for Brij® type surfactants in a previous study [13]. For example, samples produced without surfactant showed a photocatalytic half-life for the degradation of resazurin ink ranging from 9.5 to 16.5 min for the three-layer and two-layer sample respectively. Those samples produced with Tween® surfactant exhibited half-lives ranging from 3 to 5 min for two layer samples. This decrease in half-life is attributable to the reduced average particle size, as well as the increased surface roughness of surfactant enhanced thin films as shown in Table 2, which both result in an overall increased surface area to volume ratio. This enables better adsorption of the resazurin dye to the thin film surface, and a greater surface area upon which the photocatalytic reaction can occur.

Figure 3. SEM Images of samples prepared by sol-gel with decreasing concentration of Tween® 20 surfactant. D = 0.0006 mol·dm^{-3}; E = 0.0003 mol·dm^{-3}; numbers represent number of layers. Samples were annealed at 500 °C/15 min.

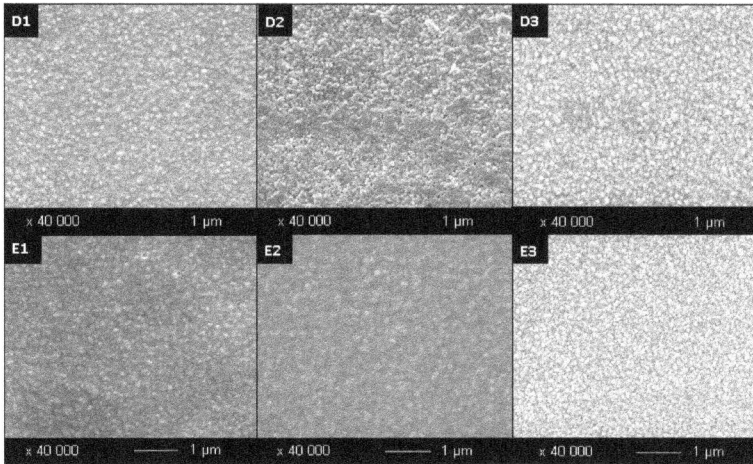

3.2. Influence of Surfactant Addition on Average Surface Roughness of Thin Films

Surfactant addition has been found to increase the root mean square surface roughness of thin films by up to 180 nm, as sample A2 (produced without surfactant) has an average root mean square surface roughness of 11 nm, compared with sample B2 (produced with 0.006 mol·dm^{-3} of Tween® 40) which has an average surface roughness of 196 nm. This large increase in surface roughness is a result of the morphological changes within the thin film that have been described, whereby the particles produced with surfactant are smaller and also more angular in shape due to the surfactant obstruction during the sol-gel growth phase.

In addition, when the concentration of the surfactant is decreased from 6×10^{-4} mol·dm^{-3} to 4×10^{-4} mol·dm^{-3}, as in samples D to E, the root mean square surface roughness decreases, whereby sample D2 exhibits the highest surface roughness of 350 nm, and E2 has a surface roughness of 191 nm. Further root mean square surface roughness values are given in Table 2. This increased surface roughness in the samples with increased concentration of surfactant can be explained by the effect of the surfactant as it surrounds the titania particles during the growth phase. When the surfactant concentration is reduced, as from sample D2 to E2, the growing particles are less restricted in their growth, meaning they can grow larger and more spherical, as can be seen in the SEM images D2 to C2 in Figure 3. This is reflected in the particle sizes, whereby sample D2 has an average particle size of 25 nm and E2 has an average particle size of 35 nm. However, it should be noted the role of agglomeration between particles present in sample D2, which also has had an effect to increase the surface roughness of the thin film. Those samples produced without surfactant have a reduced root mean square surface roughness in the range of 9–17 nm. There is no significant change in surface roughness depending on which surfactant type is used, however generally it is found that an increased concentration of surfactant increases the surface roughness. A 3D representation of the thin film surface roughness of samples produced with different surfactant types and concentrations are shown in Figure 4.

Figure 4. AFM 3D representation of thin film surface. (**a**) A2 (no surfactant); (**b**) B2 (Tween® 40, 0.006 mol·dm^{-3}); (**c**) C2 (Tween® 40, 0.003 mol·dm^{-3}); (**d**) D2 (Tween® 20, 0.006 mol·dm^{-3}); (**e**) E2 (Tween® 20, 0.003 mol·dm^{-3}).

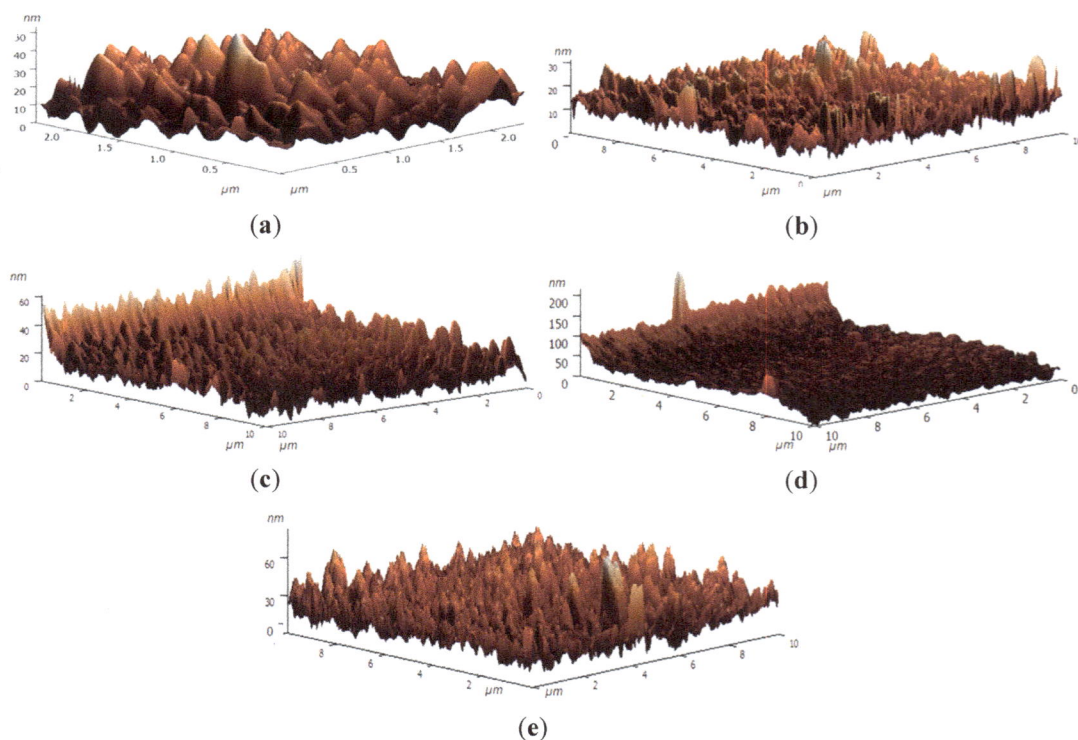

3.3. Influence of Number of Layers on Thin Film Morphology

As previously discussed, by increasing the number of consecutive dips and thereby the number of layers within the TiO_2 thin film samples, the average thickness was increased as expected. It has been found that by increasing the number of layers within TiO_2 sol-gel, the crystallite size decreases, as the individual single layer thickness increases, owing to the mechanism of growth as additional layers adhere to previous layers [22]. The addition of surfactant produces individual thicker layers, due to the increased viscosity of the sol as the surfactant coordinates around titania particles. As a result, the samples produced with surfactant are on average much thicker, with the increased concentrations of surfactant producing thicker. The addition of surfactant also made individual layers thicker, due to the increased viscosity within the thin film aiding the "sticking" of subsequent layers to the initial layers, which in turn resulted in thicker films overall. Thickness measurements were made using side-on SEM images, and are listed in Table 2. For example, sample A1 produced without surfactant and one layer, has a thickness of 42 nm, and can be compared to the one layer samples produced with addition of surfactant; B1, C1, D1 and E1 which have average thin film thicknesses of 180, 200, 110 and 120 nm respectively. In addition, as the thin film thicknesses increases with the number of layers, the particle size is also found to increase, whereby the smaller crystallites formed in the increasingly thicker layers bind together to form agglomerated particles, as specifically observed in samples C3 and D3, whereby the average particle size is 46 and 40 nm respectively. Agglomeration is also found to occur between the TiO_2 layers, whereby the initial layers can act as particle nucleation sites, for example sample B3 shows increased agglomeration owing to the higher concentration of Tween® 40 surfactant used, as seen in Figure 2.

The effect of number of layers on the root mean square surface roughness of the thin films does not have a consistent trend between samples, as seen in Table 2. Typically increasing the number of layers would be expected to increase the surface roughness of samples due to irregular adhesion between layers where different sized particles are placed on top of one another. However, most samples show a decrease in surface roughness from one layer to three layers, e.g. Sample B1 (318 nm), sample B2 (196 nm), and sample B3 (209 nm). This is likely to be the result of particles agglomerating between layers as they adhere to one another, leading to a smoother top surface overall.

3.4. Crystalline Phase Identification of TiO_2 Thin Films

A typical XRD diffraction pattern of the TiO_2 thin films deposited on $F:SnO_2$ coated glass substrate is shown in Figure 5. The thin films deposited were thinner than the $F:SnO_2$ layer (~400 nm) so breakthrough to the substrate was observed for all samples. The samples showed peaks representing a mixture of anatase, the preferred crystal phase in the [101] plane, as well as rutile, which is the more thermodynamically stable phase and was present in the [211] plane. The presence of the brookite phase was also observed in the [121], [221] and [203] planes. This mixture of phases is commonly observed in the production of TiO_2 derived thin films [2,8], and it is believed that the sol-gel method described herein has resulted in the production of largely amorphous TiO_2 thin films which cannot be detected by XRD. For further determination of the TiO_2 phase, Raman spectroscopy was used, whereby a typical spectrum is shown in Figure 6. All samples gave strong Raman bands centred at 147, 395, 513 and 642 cm^{-1}, with a weaker band centred at 198 cm^{-1}, all of which are attributable to anatase titanium dioxide indicating that whilst there is poor long range order, anatase predominates over a short range [23].

Figure 5. XRD pattern for sample B (3 layers) produced with 6×10^{-4} mol·dm^{-3} Tween® 40 surfactant and annealed at 500 °C/ 15min. Red assigned peak denotes presence of anatase in the [101] plane. Blue peaks denote presence of brookite, and black peak denotes presence of rutile. Peaks denoted with an asterisk are from the casserite substrate. This diffraction pattern was the same for samples A, C, D and E.

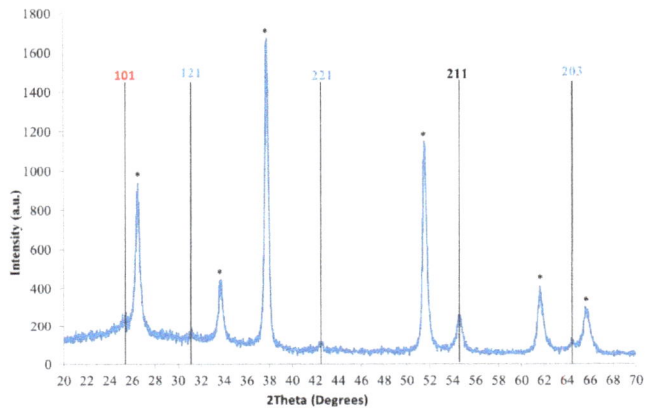

Figure 6. Typical Raman spectrum obtained for all samples. This spectra was for sample B3 annealed at 500 °C for 15 min. Peaks match the reference spectra for TiO$_2$ anatase [23].

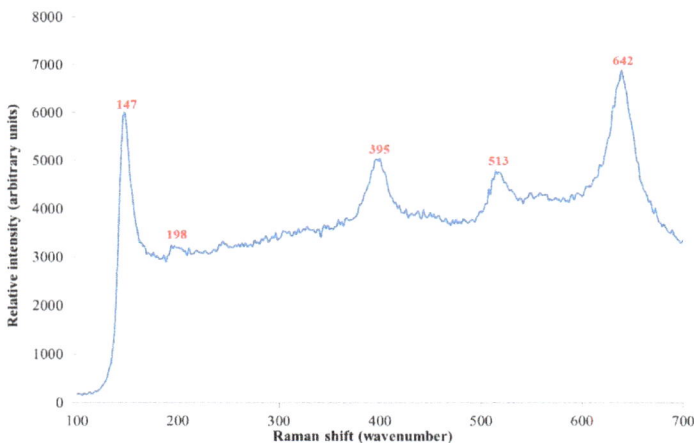

3.5. Wetting Behaviour of Thin Films

All TiO₂ thin film samples produced exhibited UV induced hydrophilicity in varying degrees, as seen by the reduction in contact angle of deionised water on the TiO₂ surface after 30 min UV irradiation shown in Table 2. Some samples that were produced with surfactant exhibited superhydrophilicity whereby the contact angle dropped to below 5° after UV irradiation, such as sample D2 and E2, which both had contact angles after 30 min UV irradiation which were around 3°. As these samples were produced with the lighter weight surfactant, Tween® 20, the effects on particle growth were that slightly larger and more spherical particles were grown in comparison to samples B and C, as the surfactant was not as bulky when surrounding titania particles during growth, therefore were not as obstructing. This can be seen in the SEM images and is also reflected in the particle sizes as shown in Table 2.

3.6. Photocatalytic Properties of Films

The sol-gel derived thin films all demonstrated photocatalytic activity for the degradation of resazurin intelligent ink. The addition of surfactant was found to increase the photocatalytic activity of the thin films as shown in Table 2 and Figure 7. The samples prepared without surfactant, samples A1–A3 exhibited half-lives in the range of 9.5–16.50 min. In contrast, the samples produced with surfactant had half-lives ranging from 3 to 5 min. The fastest half-lives observed were for samples C2 and E2, which both had a half-life of 3 min for resazurin degradation. This indicates that the addition of Tween® 40 or Tween® 20 surfactant in the lowest concentration (0.003 mol·dm^{-3}) has a beneficial effect on the thin film microstructure and morphology, such that the functional properties are improved for photocatalytic activity. The photocatalytic activity is attributed to the surfactant role as the particles grow within the sol-gel. The surfactant acts as a spacer between growing titania particles that enables great control over their size and shape, preventing agglomeration. The particles produced as a result are angular and smaller in size (average size for C2 29 nm, compared to 130 nm for A2). This modified morphology results in an increased surface area to volume ratio upon which the organic dye can be adsorbed and photocatalytically degraded.

It has been widely acknowledged that photocatalytic activity can be influenced and enhanced by a number of factors, and within this study it has been found that a combination of factors, particularly morphology and surface roughness have caused significant changes to the photocatalytic properties of the thin films produced. For example, all samples produced with surfactant exhibit a much higher average surface roughness, which can be seen morphologically in the SEM images (Figures 2 and 3) where particles are more angular and a variety of sizes, and also in the 3D AFM images in Figure 4. Samples prepared without surfactant showed a surface roughness ranging from 9 to17 nm, attributable to the poor definition of the larger, more spherical particles ranging in size 40–130 nm. In contrast, samples prepared with Tween® surfactants showed average surface roughness values ranging from 135 to 366 nm, with much more angular and variable sized particles ranging from 25 to 55 nm as their average size. The increased surface roughness enables an increased surface area to volume ratio, and as described enables greater absorption of the dye on the surface, and an increased area for the photocatalytic reaction to take place.

Figure 7. Normalised decrease in absorption of resazurin peak at 630 nm with UV irradiation (365 nm) over time for samples: Blank; A = no surfactant; B = Tween® 40, 0.006 mol·dm^{-3}; C = Tween® 40, 0.003 mol·dm^{-3}; E = Tween® 20, 0.003 mol·dm^{-3}. Sample D has been removed for graph clarity.

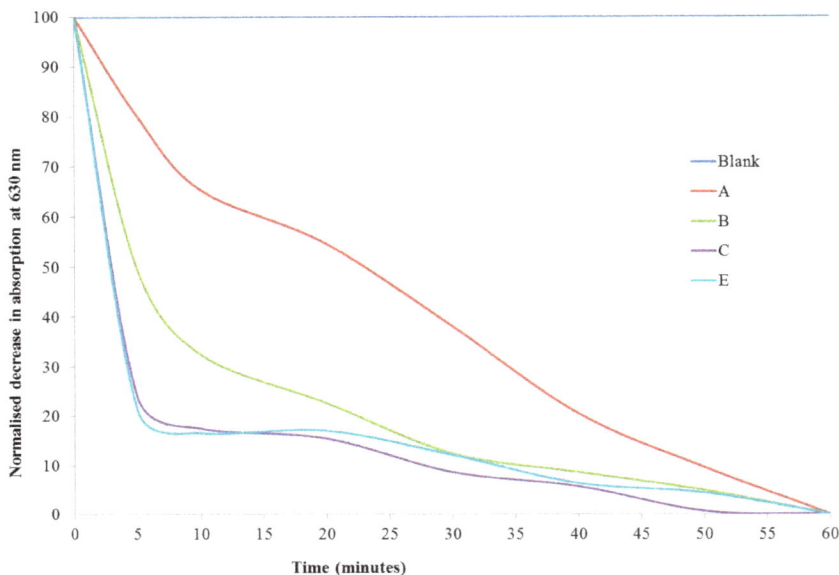

4. Conclusions

TiO$_2$ thin films which were photocatalytically active were deposited using a non-aqueous modified sol gel method with the addition of two Tween® type surfactants at different concentrations. The addition of surfactant was found to reduce the photocatalytic half-life of the reazurin dye from 16.5 to 3 min for a sample deposited from a sol that contained Tween® 40 or 20 surfactant in a concentration of 3×10^{-4} mol·dm^{-3}. The improved photocatalytic activity is attributed to the angular shape of particles, as well as a smaller average particle size range and increased root mean square surface roughness, indicating a greater surface area for catalysis to occur upon.

References

1. Çamurlu, H.E.; Kesmez, Ö.; Burunkaya, E.; Kiraz, N.; Yeşil, Z.; Asiltürk, M.; Arpaç, E. Sol-gel thin films with anti-reflective and self-cleaning properties. *Chem. Pap.* **2012**, *66*, 461–471.
2. Page, K.; Palgrave, R.G.; Parkin, I.P.; Wilson, M.; Savin, S.L.P.; Chadwick, A.V. Titania and silver–titania composite films on glass–potent antimicrobial coatings. *J. Mater. Chem.* **2007**, *17*, 95–104.
3. Ni, M.; Leung, M.K.H.; Leung, D.Y.C.; Sumathy, K. A review and recent developments in photocatalytic water-splitting using TiO$_2$ for hydrogen production. *Renew. Sustain. Energy Rev.* **2007**, *11*, 401–425.

4. Macwan, D.P.; Dave, P.N.; Chaturvedi, S. A review on nano-TiO$_2$ sol–gel type syntheses and its applications. *J. Mater. Sci.* **2011**, *46*, 3669–3686.

5. Kim, D.S.; Kwak, S.-Y. The hydrothermal synthesis of mesoporous TiO$_2$ with high crystallinity, thermal stability, large surface area, and enhanced photocatalytic activity. *Appl. Catal. A Gen.* **2007**, *323*, 110–118.

6. Shinde, P.S.; Bhosale, C.H. Properties of chemical vapour deposited nanocrystalline TiO$_2$ thin films and their use in dye-sensitized solar cells. *J. Anal. Appl. Pyrolysis.* **2008**, *82*, 83–88.

7. Choy, K.L. Chemical vapour deposition of coatings. *Prog. Mater. Sci.* **2003**, *48*, 57–170.

8. Edusi, C.; Hyett, G.; Sankar, G.; Parkin, I.P. Aerosol-assisted CVD of titanium dioxide thin films from methanolic solutions of titanium tetraisopropoxide; substrate and aerosol-selective deposition of rutile or anatase. *Chem. Vap. Depos.* **2011**, *17*, 30–36.

9. Romero, L.; Binions, R. Effect of AC electric fields on the aerosol assisted chemical vapour deposition growth of titanium dioxide thin films. *Surf. Coat. Technol.* **2013**, *230*, 196–201.

10. Arconada, N.; Durán, A.; Suárez, S.; Portela, R.; Coronado, J.M.; Sánchez, B.; Castro, Y. Synthesis and photocatalytic properties of dense and porous TiO$_2$-anatase thin films prepared by sol–gel. *Appl. Catal. B Environ.* **2009**, *86*, 1–7.

11. Yang, P.; Zhao, D.; Margolese, D.I.; Chmelka, B.F.; Stucky, G.D. Block copolymer templating syntheses of mesoporous metal oxides with large ordering lengths and semicrystalline framework. *Chem. Mater.* **1999**, *11*, 2813–2826.

12. Stathatos, E.; Lianos, P.; Tsakiroglou, C. Highly efficient nanocrystalline titania films made from organic/inorganic nanocomposite gels. *Microporous Mesoporous Mater.* **2004**, *75*, 255–260.

13. Anderson, A.-L.; Binions, R. The effect of Brij® surfactants in sol–gel processing for the production of TiO$_2$ thin films. *Polyhedron* **2015**, *85*, 83–92.

14. Yang, J.; Peterlik, H.; Lomoschitz, M.; Schubert, U. Preparation of mesoporous titania by surfactant-assisted sol–gel processing of acetaldoxime-modified titanium alkoxides. *J. Non. Cryst. Solids.* **2010**, *356*, 1217–1227.

15. Uchida, H.; Patel, M.N.; May, R.A.; Gupta, G.; Stevenson, K.J.; Johnston, K.P. Highly-ordered mesoporous titania thin films prepared via surfactant assembly on conductive indium–tin-oxide/glass substrate and its optical properties. *Thin Solid Films* **2010**, *518*, 3169–3176.

16. Černigoj, U.; Štangar, U.L.; Trebše, P.; Krašovec, U.O.; Gross, S. Photocatalytically active TiO$_2$ thin films produced by surfactant-assisted sol–gel processing. *Thin Solid Films* **2006**, *495*, 327–332.

17. Pelentridou, K.; Stathatos, E.; Lianos, P.; Drakopoulos, V. A new precursor for the preparation of nanocrystalline TiO$_2$ films and their photocatalytic properties. *J. Nanosci. Nanotechnol.* **2010**, *10*, 6093–6098.

18. Choi, H.; Stathatos, E.; Dionysiou, D.D. Synthesis of nanocrystalline photocatalytic TiO$_2$ thin films and particles using sol–gel method modified with nonionic surfactants. *Thin Solid Films.* **2006**, *510*, 107–114.

19. Schneider, C.A.; Rasband, W.S.; Eliceiri, K.W. NIH Image to ImageJ: 25 years of image analysis. *Nat. Methods.* **2012**, *9*, 671–675.

20. Tauc, J. Optical properties and electronic structure of amorphous Ge and Si. *Mater. Res. Bull.* **1968**, *3*, 37–46.

21. Kafizas, A.; Mills, A.; Parkin, I.P. A comprehensive aerosol spray method for the rapid photocatalytic grid area analysis of semiconductor photocatalyst thin films. *Anal. Chim. Acta* **2010**, *663*, 69–76.

22. Wang, C.; Meinhardt, J.; Löbmann, P. Growth mechanism of Nb-doped TiO_2 sol–gel multilayer films characterized by SEM and focus/defocus TEM. *J. Sol-Gel Sci. Technol.* **2009**, *53*, 148–153.

23. Ohsaka, T.; Izumi, F.; Fujiki, Y. Raman spectrum of anatase, TiO_2. *J. Raman Spectrosc.* **1978**, *7*, 321–324.

Photocatalytic Activity and Stability of Porous Polycrystalline ZnO Thin-Films Grown via a Two-Step Thermal Oxidation Process

James C. Moore, Robert Louder and Cody V. Thompson

Abstract: The photocatalytic activity and stability of thin, polycrystalline ZnO films was studied. The oxidative degradation of organic compounds at the ZnO surface results from the ultraviolet (UV) photo-induced creation of highly oxidizing holes and reducing electrons, which combine with surface water to form hydroxyl radicals and reactive oxygen species. Therefore, the efficiency of the electron-hole pair formation is of critical importance for self-cleaning and antimicrobial applications with these metal-oxide catalyst systems. In this study, ZnO thin films were fabricated on sapphire substrates via direct current sputter deposition of Zn-metal films followed by thermal oxidation at several annealing temperatures (300–1200 °C). Due to the ease with which they can be recovered, stabilized films are preferable to nanoparticles or colloidal suspensions for some applications. Characterization of the resulting ZnO thin films through atomic force microscopy and photoluminescence indicated that decreasing annealing temperature leads to smaller crystal grain size and increased UV excitonic emission. The photocatalytic activities were characterized by UV-visible absorption measurements of Rhodamine B dye concentrations. The films oxidized at lower annealing temperatures exhibited higher photocatalytic activity, which is attributed to the increased optical quality. Photocatalytic activity was also found to depend on film thickness, with lower activity observed for thinner films. Decreasing activity with use was found to be the result of decreasing film thickness due to surface etching.

Reprinted from *Coatings*. Cite as: Moore, J.C.; Louder, R.; Thompson, C.V. Photocatalytic Activity and Stability of Porous Polycrystalline ZnO Thin-Films Grown via a Two-Step Thermal Oxidation Process. *Coatings* **2014**, *4*, 651–669.

1. Introduction

Zinc oxide (ZnO) is a highly useful and practical wide bandgap semiconducting material with a broad range of applications, including self-cleaning and anti-fogging surfaces, sterilization, gas sensing, energy production and environmental purification [1–4]. Specifically, ZnO efficiently absorbs ultraviolet (UV) light and has surface electrical properties sensitive to the environment at the interface, with device applications that include gas sensors, photovoltaic cells, light emitting diodes and photocatalysts [1,5–10]. The photocatalytic effects of ZnO are being exploited for use within self-cleaning paints, in environmental remediation applications and prophylactics with nanoparticle and colloidal suspensions demonstrating high photodegradation efficiency for organic compounds [11]. Thin films have received recent interest due to their reusability and transparency, which is essential for applications, such as self-cleaning glass and antimicrobial coatings on solid surfaces and flexible plastics [12–14]. Transparent ZnO films could also find use as fingerprint-resistant barriers on touch screen devices, such as cell phones and tablet computers.

Many environmental pollutants are organic in nature, and many proposed methods of environmental decontamination involve oxidation of the organic pollutants [15]. However, using semiconductor photocatalysts to oxidize and remove such pollutants from the local environment has many advantages over alternative methods [16]. ZnO materials in particular are nontoxic and present little additional harm to the environment in which they are used, contrary to most other methods of decontamination. However, there is some concern about the dissolution of ZnO particles and resulting Zn toxicity in marine environments [4,17]. Furthermore, ZnO photocatalysts do not need to be re-activated after undergoing photoinduced oxidation and reduction reactions. Conversely, activated carbon, a popular choice for water purification, requires expensive and potentially polluting reactivation [18].

Another traditional means of decontamination involves microorganisms, such as bacteria, which biologically degrade toxic organics [18]. However, these processes occur at a much slower rate compared to photocatalytic oxidation by semiconductors, such as ZnO, and are inefficient at concentrations below parts-per-million (ppm) levels, while ZnO photocatalysts have been shown to oxidize pollutants present in extremely low concentrations. Additionally, many pollutants can also be toxic to the microorganisms themselves, reducing their catalytic activity with time. ZnO photocatalysts degrade most organic pollutants non-selectively, though their stability is a topic of study [19].

Extensive work has gone into investigating the photocatalytic properties of ZnO nanoparticles and colloid suspensions [20,21]. For environmental remediation purposes, nanoparticle powders are particularly effective, since they can be readily mixed with the contaminated solution and have a high surface area. However, separating the catalyst from solution is challenging, which makes their use in these applications potentially cost-prohibitive [22]. As mentioned, Zn toxicity is also a concern for these systems, especially if allowed to remain in the environment. Fujishima *et al.* suggest that the nanofilm form of these semiconductors is preferable to particles for use in fluid decontamination exactly because the nanoparticles need to eventually be collected and removed from the fluid [23,24].

There has been little discussion in the literature concerning the sorts of structural and photophysical properties that can directly affect photocatalytic activity for film-based catalysts. As an example, whether degradation occurs predominantly due to reaction with free-radicals or directly with the holes themselves is controversial, with some groups even proposing a predominant electron-based catalytic pathway on ZnO single-crystal surfaces [23,25]. For surface-based applications, challenges exist with nanoparticle-based films, such as adhesion and optical transparency [26]. Nanoparticles do have a high surface area per volume, which increases the number of available surface states to serve as reaction sites. However, increased crystallinity associated with larger particle sizes typically results in greater optical efficiency and, therefore, higher electron-hole pair production efficiency [23,27,28]. Many applications of decontamination using semiconductor photocatalysts involve the Sun as a practical source of UV illumination, although only 2%–3% of solar radiation will induce semiconductor-catalyzed oxidation [29]. Accordingly, the photodegradation efficiencies of the semiconductor photocatalysts designed for these uses need to be optimized in order to be of practical use. Both high surface area and optical efficiency are required for high photocatalytic activity with metal-oxides. For TiO_2 and ZnO thin, polycrystalline films, a decrease in grain size typically results in increased surface roughness and surface area; however, small grain size typically also corresponds to an increase in deep-level defects that lower the number of photo-induced holes at the surface available for catalysis [28]. These competing mechanisms must be balanced.

In previous work, we have found that thin (<200 nm) ZnO films grown via thermal oxidation of Zn-metal at relatively low temperature (300 °C) result in high surface roughness with low deep-level defects [14,28]. Increasing surface-level Zn interstitials could also result in greater catalytic activity and a favorable shift in wavelength into the visible spectrum. We have also found that significant blue emission associated with Zn interstitials near the surface and very little deep-level emission from bulk-related defects can be obtained via tailoring of the films thickness and grain size, resulting in a potential increase in photocatalytic activity due to a favorable balance of features [28,30].

In this study, we measure the photocatalytic activity and stability of thin, polycrystalline ZnO films fabricated on sapphire substrates via direct current (DC) sputter deposition of Zn-metal films, followed by thermal oxidation at several annealing temperatures. In particular, we describe growth parameters that result in highly porous, polycrystalline films demonstrating high surface roughness while simultaneously exhibiting a high excitonic-to-green emission ratio. We also investigate the time-dependent stability of these films.

In Section 2, we describe the process by which films have been fabricated and characterized, and we discuss the method used for determining catalytic activity. In Section 3, we discuss the resulting morphological, structural and optical properties of the fabricated films. In Section 4, we discuss surface catalysis pathways and the reaction kinetic models used to describe the catalytic activity of these films. Finally, in Section 5, we discuss the balance between crystal grain size and the optical efficiency that affects the photocatalytic activity of polycrystalline ZnO films.

2. Experiment

Zinc films where deposited on *c*-plane sapphire substrates via direct current sputter deposition. Metallic Zn targets were obtained commercially and had a purity of 99.99%. Before deposition, substrates were cleaned via immersion in acetone, ultrasonically cleaned in methanol and rinsed in deionized water. The chamber base pressure was maintained between 1.0×10^{-5} and 2.5×10^{-5} mbars. A gate valve between the chamber and the pump was utilized as a throttle to maintain an Ar pressure of approximately 2×10^{-2} mbars. Sputtering power was maintained between 10 and 20 W with the substrate located approximately 10 cm from the sputter source. Deposition times ranged from 15 to 40 min, resulting in film thicknesses between 100 nm and 600 nm as measured via the AFM profile and reflectometry. Thermal oxidation of the Zn metal films was carried out in an air-ambient muffle furnace. For all samples, Zn films were initially annealed at 300 °C for 9–24 hours to ensure complete oxidation. Some films were then re-annealed for 1 hour at 600 °C, 900 °C and 1200 °C.

Zinc oxide films were characterized via X-ray diffraction (XRD), atomic force microscopy (AFM) and photoluminescence (PL). Structural properties of the ZnO films were measured using Cu-Kα radiation in the range from 30° to 50°. The morphology of the films was determined via dynamic-mode AFM using an Anfatec Level AFM and approximately 300 kHz resonant aluminum backside silicon tips. Photoluminescence spectra were obtained at room temperature using a HeCd laser as an excitation source and a power of $P = 0.3$ W/cm^2. UV illumination was provided by a deuterium lamp.

Following morphological and optical characterization, the photocatalytic activities of the films were characterized by measuring the degradation of Rhodamine B dye (rhoB) in solution. RhoB was used to simulate an organic environmental contaminant, because its concentration in solution can be accurately measured spectrophotometrically. Oxidized rhoB products do not absorb visible light, so the concentration and the resulting absorbance of the rhoB solutions decrease as the rhoB is oxidized by the ZnO photocatalysts. The overall photocatalytic activities of the materials therefore correlate to the rate of change in the concentration of the rhoB solutions as measured by UV-Vis spectrophotometry.

The UV-Vis spectrophotometer was first calibrated in order to determine the relationship between absorbance measurements and rhoB concentrations. This was accomplished by measuring the absorbance of known concentrations of rhoB and constructing a calibration curve from the resulting measurements (not shown). The concentrations in the rest of the experiment were then calculated from the absorbance measurements. The linear range of the absorbance-concentration relationship was determined to be between 1 and 8 ppm rhoB, so an initial concentration of 8 ppm rhoB in solution was used for the photodegradation experiments.

Each ZnO thin film was incubated in an 8 ppm rhoB solution while irradiated with UV light at a constant power density of 80 μW/cm. The absorbance of the solution was taken after 10, 20, 30, 45, 60, 90, 120, 150 and 180 min of UV irradiation at the peak absorbance wavelength (553 nm) for maximum sensitivity, and the corresponding rhoB concentration was calculated using the equation of the rhoB calibration curve. In order to account for differences in the surface areas of the films, the total change in concentration of rhoB at each time interval was divided by the specific surface area of the film and plotted against the time of UV irradiation.

3. Film Growth and Characterization

As described earlier, greater catalytic activity is expected with high surface roughness and high optical efficiency. In this section, we discuss the morphology, structure and optical properties of the ZnO thin films under study. Specifically, we describe the morphological and structural evolution with annealing temperature and film thickness, concentrating on crystal grain sizes and surface roughness. We also describe how growth parameters affect the optical properties of the films. In particular, we discuss the relative excitonic emission efficiency and yellow-green band emission.

3.1. Morphology and Structure

The evolution of the surface morphology with increasing annealing temperature for 200 nm- and 600 nm-thick ZnO films is shown in the AFM images presented in Figure 1a–e. As shown in Figure 1a, the as-grown zinc film demonstrates a high surface roughness and approximately 100-nm diameter protrusions. Figure 1b,c shows the surface morphology of resulting 600 nm-thick ZnO films annealed at 300 °C and 600 °C. Figure 1d,e shows the surface morphology of the resulting 200 nm-thick ZnO films annealed at 300 °C and 600 °C. For both thicknesses, there is very little change in the underlying characteristics between zinc-metallic films and ZnO films annealed at 300 °C; however, surface roughness is observed to increase, and protrusions grow in size by approximately 50 nm. This is consistent with previous studies, where Gupta *et al.* show that the preferred orientation of ZnO thermally oxidized on glass can depend on the Zn film texture and oxidizing agent [31]. For 600 nm-thick films, an increase in surface roughness is observed with increasing temperature (Figure 1b,c). Interestingly, at temperatures above 600 °C and a thickness of 600 nm, long vertically-aligned nanorods are visible, which is consistent with previous studies [32]. For a 200 nm-thick films, a decrease in protrusion diameter and surface roughness is observed with increasing temperature (Figure 1d,e). Nanorods are not seen for the thinner films annealed at any temperature. For both film thicknesses, there is little change in surface morphology observed at higher temperatures (not shown). Specifically, the protrusion size does not significantly change.

Grain size was characterized by both AFM and XRD. For films having thicknesses between 400 nm and 600 nm, grain size was observed to increase with increasing annealing temperature from 300 °C to 1200 °C. Interestingly, for films having thicknesses between 100 nm and 200 nm, grain size decreased with increasing temperature up to a certain point. Figure 1d,e shows a decrease in the protrusion diameter from approximately 150 nm to approximately 100 nm at annealing temperatures of 300 °C and 600 °C, respectively, with no further significant change in size as the temperature was further increased. Grain size, as determined by the full width at half maximum in the XRD spectra (not shown), was also found to decrease with increasing temperature up to 600 °C, with relatively small increases at higher temperatures, which is consistent with AFM measurements. This observation appears inconsistent with some reports in the literature for thermally-oxidized ZnO films [33,34]. This discrepancy may be the result of differences in studied temperature regimes, the variations in film thickness and/or our two-step thermal annealing process, where metallic zinc films are all initially oxidized at low temperature. Furthermore, our metallic zinc films display a significantly different texture and larger initial particle size in comparison to films grown via other methods, which has been shown to affect resulting film morphology and structure [31].

3.2. Optical Properties

Figure 2 shows the PL spectra of ZnO films with a thickness of (a) 600 nm and (b) 200 nm thermally annealed at various temperatures. The spectra for the 600 nm-thick films all show asymmetric broad bands in the yellow-green region, with no significant yellow-green emission observed for the 200 nm-thick films. For all thicknesses, a more narrow band in the UV is observed, which is associated with excitonic emission; however, the 200-nm film demonstrates an asymmetric and broad band in the blue-UV region at higher temperatures. For all thicknesses, the strongest UV excitonic emission is observed at low temperature, with decreasing UV emission observed with increasing annealing temperature. Interestingly, for thinner films, increasing temperature results in a significant redshift (0.15 eV) in the UV excitonic peak and an asymmetrical peak broadening. This change in UV peak position could be explained by a corresponding shift in bandgap energy, which would be consistent with transmission studies of ZnO films grown via the sol-gel method and previous studies of thermally annealed Zn films [28,35]. Wu *et al.* and Dijken *et al.* both demonstrate that an increase in particle size should result in a redshift in energies, which appears inconsistent with our results, since in this temperature and thickness regime, we see a decrease in grain size [36,37]. However, these studies discuss systems where quantum size effects become relevant, and the particle sizes in this study are sufficiently large, such that the shift in bandgap cannot be explained via a similar mechanism. Jain *et al.* speculate that this red shift is the result of an increase in interstitial zinc atoms, which we demonstrated to be the case in a previous study [28,35].

Figure 1. AFM topography images of (**a**) Zn-metal films before oxidation (grayscale range = 200 nm) and the resulting polycrystalline ZnO films after annealing in ambient air; Six hundred nanometer-thick films annealed at (**b**) 300 °C (grayscale range = 500 nm) and (**c**) 600 °C (grayscale range = 1000 nm). Two hundred nanometer-thick films annealed at (**d**) 300 °C (grayscale range = 300 nm) and (**e**) 600 °C (grayscale range = 250 nm).

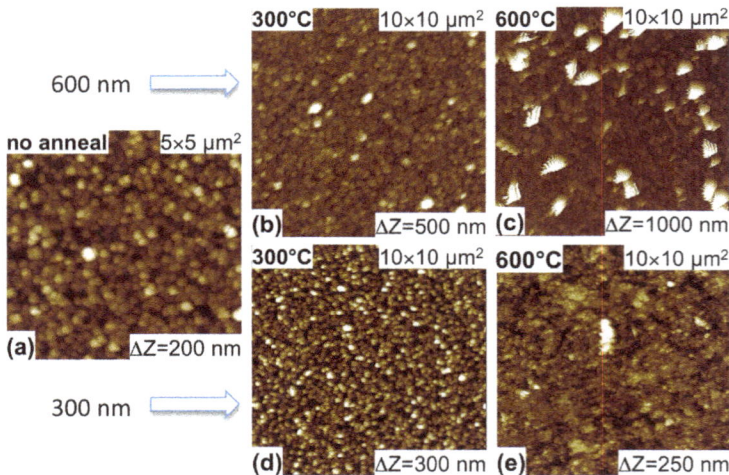

As shown in Figure 2, 600 nm-thick films demonstrate increasing yellow-green band intensity with increasing annealing temperature when compared to UV emission. This can be attributed to a rapid increase in V_o^+ and O_i^- ion centers at high temperatures [34,36]. In contrast, 200 nm-thick films demonstrate little green and yellow band emission at any temperature (Figure 2b). Deep level emission is attributed to bulk defects; therefore, it is possible that decreased bulk volume results in the formation of relatively fewer deep-level states. If green and yellow emission results from the recombination of a delocalized electron close to the conduction band with a deeply trapped hole in the V_o^+ and O_i^- centers in the bulk, respectively, then a decrease in film thickness would decrease the bulk with respect to the depletion region, resulting in weaker bulk-related, deep-level emission [36]. Reaction kinetics could also contribute, with thinner films having shorter diffusion paths for reactive oxygen species during oxidation [38]. Thin films would therefore demonstrate fewer V_o^+ and O_i^- ions at any annealing temperature, as observed.

Figure 2. PL spectra of ZnO films grown at 300 °C, 600 °C, 900 °C and 1200 °C at thicknesses of (**a**) 600 nm and (**b**) 200 nm. For thicker films, increasing green band emission relative to UV emission is seen with increasing temperature. Thinner films demonstrate little green band emission; however, a significant redshift in UV emission and asymmetrical band broadening is observed at higher temperatures.

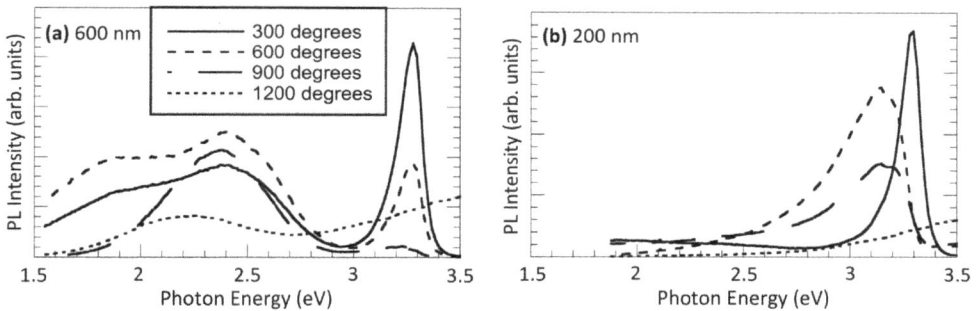

For thin films at high temperatures, the asymmetric and broad UV excitonic emission bands result from increasing blue emission and corresponding decreasing UV emission [28]. As shown in Figure 2b, the PL spectra for the 200 nm-thick film exhibited the most dramatic blue band emission and very low green band emission at all temperatures. Wang *et al.* found that the intensities of the green and yellow cathodoluminescence peaks were strongly affected by the width of the free-carrier depletion region near the surface [34]. They argue that single ionized oxygen vacancies exist only in the bulk, so the magnitude of the depletion region in relation to the bulk directly affects the intensity of the green emission in the cathodoluminescence spectrum. It has been suggested that blue emission results from zinc interstitials found in the depletion region, so an approximately 400-nm blue emission should only be observed in a sample with a wide depletion region relative to the bulk. Otherwise, deep-level green emission will dominate [28,39,40]. Therefore, the emergence of blue emission with decreasing film thickness results from a low ratio of the bulk to the depletion region.

Temperature and grain size also contribute to the PL spectra, since no blue emission is observed for 200-nm films annealed at 300 °C (see Figure 2b). As discussed, we observe a larger grain size via XRD and AFM for 200-nm films annealed at 300 °C, resulting in a larger bulk to depletion region ratio, which could contribute to the weaker blue emission. Furthermore, thinner films have shorter diffusion paths for reactive oxygen species during oxidation, and lower temperatures slow the Zn/O$_2$ reaction. This results in fewer zinc interstitials, which would manifest as weaker blue emission and a red shift in both blue and UV emissions, as is observed.

For photocatalysis applications, thin ZnO films show significant promise due to their potential for balancing high optical quality with high surface roughness. In particular, 200 nm-thick films annealed at low temperature (300 °C) demonstrate a high UV-to-green emission ratio with relatively high surface roughness. Low-temperature annealed 600-nm thick films also show strong excitonic photoemission combined with a tall protrusion height within the porous film structure, which could result in greater effective surface area for photochemical reactions. In this study, we investigate photocatalysis with these thin films, because they show the most promise for high photocatalytic activity.

4. Surface Catalysis and Reaction Kinetics

Metal-oxide semiconductors use light energy to catalyze oxidation-reduction reactions via electron-hole pair production at the material surface [41,42]. The photogenerated holes on the semiconductor surface have a high oxidation potential, while the photogenerated electrons have a high reduction potential [19,23]. Several types of aqueous reactions catalyzed by electron-hole pairs lead to the formation of the hydroxyl radicals and reactive oxygen species that are directly responsible for the oxidative degradation of organic compounds. One of these reactions involves the oxidation of water (H_2O) by a hole (h^+) into hydrogen ions (H^+) and a hydroxyl radical (O^*):

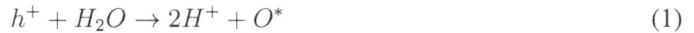

$$h^+ + H_2O \rightarrow 2H^+ + O^* \tag{1}$$

Another reaction involves the reduction of molecular oxygen (O_2) into a superoxide radical (O_2^{-*}) by a photogenerated electron (e^-):

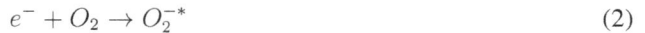

$$e^- + O_2 \rightarrow O_2^{-*} \tag{2}$$

The superoxide radical can be further reduced by another electron and then paired with two H^+ ions to form hydrogen peroxide (H_2O_2):

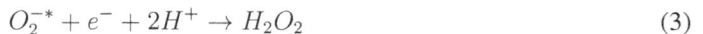

$$O_2^{-*} + e^- + 2H^+ \rightarrow H_2O_2 \tag{3}$$

The hydrogen peroxide can then be reduced by an electron to form hydroxyl radicals:

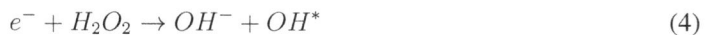

$$e^- + H_2O_2 \rightarrow OH^- + OH^* \tag{4}$$

By these reactions, either the electrons or the holes of the photogenerated electron-hole pairs can produce hydroxyl and superoxide radicals that can subsequently degrade organic compounds [19,23]. It has also been proposed that the electrons or holes themselves may be responsible for at least some of the degradation of organics. Whether degradation occurs predominantly due to reaction with free-radicals or directly with the holes themselves is controversial, with some groups even proposing a predominant electron-based catalytic pathway on ZnO single-crystal surfaces [23,25]. Most likely, however, a large combination of chemical reactions involving various intermediates ultimately leads to the degradation of the organic compounds, which complicates the mathematical descriptions of the rates at which the overall degradation occurs [19].

Simplified mathematical descriptions of the degradation reactions have been used to experimentally quantify and compare photocatalytic activities. For most purposes, especially when the concentration of the contaminant is less than 10 ppm, the reaction can be considered first order, meaning that the time t rate of change in concentration c of the contaminant, dc/dt, follows the general relationship:

$$-\frac{dc}{dt} = kc \tag{5}$$

where k is the rate constant [19]. This equation has been modified for solid photocatalysts to account for the surface area of the photocatalyst and the intensity of the incident light, as follows:

$$-\frac{dc}{dt} = kcA_S\sqrt{I} \tag{6}$$

where I is the intensity of light used and A_S is the total surface area of the photocatalyst [23]. The solution to this differential equation is as follows:

$$c(t) = c_0 e^{-ktA_S\sqrt{I}} \tag{7}$$

where c_0 is the initial concentration of the contaminant. In order to make the concentration-time relationship linear, the equation can be rewritten as the integrated rate law, as follows:

$$\ln\frac{c}{c_0} = -ktA_S\sqrt{I} \tag{8}$$

When the natural log of the normalized concentration c/c_0 is plotted as a function of time, the rate constant k can be easily determined from the slope of the linear best-fit. The rate constant is the figure of merit referred to in the literature for quantitative comparisons of photocatalytic activity [19].

Zero- and half-order rate laws have also been used to describe the photocatalyst degradation reactions. In particular, the half-order rate law has been shown to more accurately model photocatalytic reactions on metal-oxide surfaces, most likely due to the combination of zero- and first-order chemical reactions involved in photodegradation [19]. The differential and integrated forms of the half-order rate law are as follows:

$$-\frac{dc}{dt} = A_S\sqrt{I}kt^{1/2} \tag{9}$$

and:

$$c^{1/2}(t) = c_0^{1/2} - A_S\sqrt{I}\frac{k}{2}t \tag{10}$$

respectively. In this study, we calculate $(c^{1/2} - c_0^{1/2})/A_S$ and plot this as a function of the UV irradiation time, with the half-order rate constant determined from the slope of the linear best-fit. While the light intensity is a factor in calculating the rate constant, the same light intensity was used for each experiment in this study, and its effect on the rate of each reaction can be negated.

5. Results

In this section, we discuss the photocatalytic activity of the fabricated ZnO thin films and its dependency on the annealing temperature and film thickness. We also discuss the stability of these films with continued use by investigating the time-dependence of the photocatalytic activity. As described in Section 3, the annealing temperature and film thickness have been shown to affect both the surface roughness and the efficiency of electron-hole pair production, which have been shown to affect photcatalysis. Furthermore, surface degradation of ZnO with exposure to aqueous environments has been reported, which could affect the long-term stability of ZnO thin films with respect to photocatalysis [4,17].

5.1. Reaction Order

Figure 3a,b shows that concentrations of rhoB fit to the first- and half-order integrated rate law, respectively, for increasing annealing temperature and constant 200 nm-thick films. Half-order reaction kinetics (Equation (10)) better approximate the overall degradation reaction than the more commonly applied first-order rate law (Equation (8)), as indicated by the higher coefficient of determination (R^2) values for the linear fits (shown in Table 1). The orders of the individual reactions that ultimately lead to the degradation of organic compounds, such as rhoB by semiconductor photocatalysts, vary between zero and one, depending on the reaction. When these reactions are coupled, the order of the overall reaction is somewhere between zero and one, and so the best approximation is most likely to be the half-order rate law [19]. For reasons of simplicity and uniformity between studies, the first-order rate law has commonly been applied to quantify the photocatalytic activities of semiconductor photocatalysts. However, because the half-order rate law is a better and more logical approximation, the photocatalytic activities were calculated in this study using the half-order integrated rate law (Equation (10)).

Figure 3. Concentrations of rhoB fit to the (**a**) first order and (**b**) half-order integrated rate law for varying annealing temperature (300–900 °C). The best fit curves are indicated by the solid lines, with the slope of the best fit curves representing the photocatalytic activity. The control consists of degradation measurements made in the absence of catalyst films.

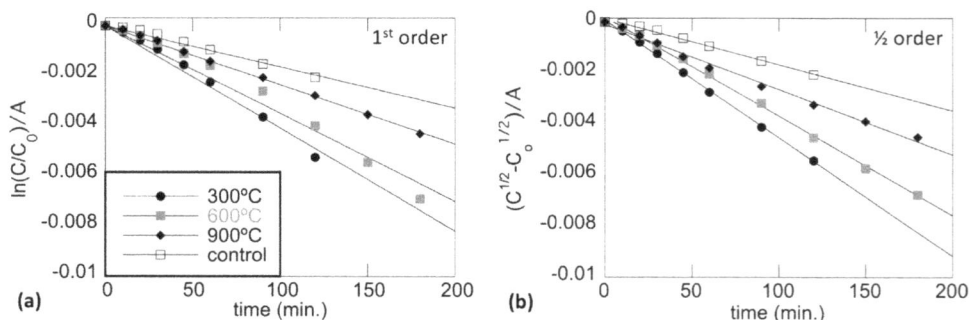

Table 1. Regression statistics for first- and half-order reaction kinetics.

Annealing Temperature (°C)	R^2 1st order	R^2 half-order
300	0.999	0.856
600	0.998	0.887
900	0.993	0.886

However, it should be noted that the order of the overall reaction is determined by which reaction types predominate at any given moment, which itself is dependent on a multitude of factors. One factor that determines the predominant reaction that is taking place at any given moment is the concentration of the various molecules involved in the degradation reactions, which varies significantly over time. The order of the overall reaction can thus change over the course of the photocatalysis experiment due to decreasing concentrations of rhoB or increasing concentrations of free radicals and peroxides in solution. The change in overall reaction order is apparent from inspection of Figure 4, which is a semi-log plot representing the first-order rate law. The semi-log concentration is linear during the first 60 min of irradiation time and significantly non-linear thereafter, suggesting a change in overall reaction order. Therefore, there is a limit to the ability of the reaction to follow any integrated rate law with time, which is not necessarily a result of the catalyst film itself. For studies involving long-term film stability, activities are measured in shorter 60–120-min time intervals before the rhoB solution is replenished.

Figure 4. Semi-log plot of rhoB concentration with time. Within the first 60 min of irradiation, first-order reaction kinetics fit the observed degradation as indicated by the solid line. However, the change in overall reaction order is apparent after the first 60 min.

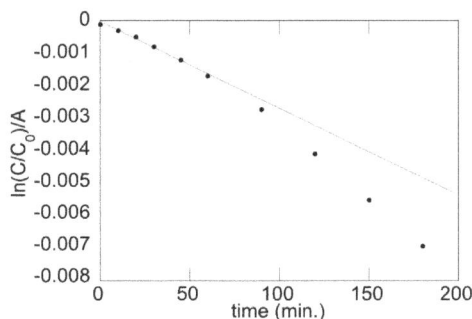

5.2. Photocatalytic Activity and Annealing Temperature

As shown in Figure 3b, the half-order photocatalytic activity decreases with increasing annealing temperature, as indicated by the decreasing slope of the best-fit lines with increasing temperature. A control measurement was made, which consisted of a sapphire substrate without a ZnO film incubated in an 8-ppm rhoB solution, while irradiated with UV light (hollow squares, Figure 3b). We observed no difference in photodegradation between rhoB solutions alone and rhoB solutions with immersed sapphire, suggesting that the substrate plays no significant role in the degradation of the rhoB. It is clear that some degradation of rhoB occurs as a result of UV illumination alone. To determine the contributions to degradation by the ZnO films and, therefore, the photocatalytic activity of the films, we subtract the control curve from measurements made in ZnO-incubated solution, with the resulting slope of the best-fit representing the photocatalytic activity of the metal-oxide films. The photocatalytic activities of 200 nm-thick ZnO films determined from half-order kinetics are reported in Table 2.

The activity of the films annealed at 600 °C was reduced by 20% compared to the films annealed at 300 °C, and the activity of the 900 °C annealed films was reduced by 30% compared to the 600 °C annealed films. Interestingly, this is the same proportional reduction as is observed for peak PL excitation and annealing temperature, as shown in Figure 2b. This suggests that excitation efficiency, specifically in the excitonic band, can significantly affect the photocatalytic activity of metal-oxide films.

Table 2. Photocatalytic activity at various annealing temperatures.

Annealing temperature (°C)	Photocatalytic activity ($ppm^{1/2}mm^{-2}min^{-1}$)
300	$(57 \pm 1) \times 10^{-6}$
600	$(31 \pm 1) \times 10^{-6}$
900	$(10 \pm 1) \times 10^{-6}$

Since PL excitation is directly correlated to electron-hole pair formation on the semiconductor surface, the films with higher optical quality have higher rates of electron-hole pair formation. Furthermore, the oxidation and reduction reactions that lead to the degradation of rhoB and other organic compounds by the semiconductor photocatalysts are dependent on the availability of electrons and holes on the surface of the semiconductor, so the optical quality of the films is the most important factor in the photocatalyzed degradation by the semiconductor, at least when morphological features are similar.

The effective surface area on the nano-scale should also affect the photocatalytic activities of the films, though the precise extent of its significance has been difficult to determine for metal-oxides due to the connection between optical properties and grain size, surface roughness and other morphological properties [43]. Since the films annealed at 300 °C in this study have larger grain sizes than those annealed at higher temperature, their surface area is effectively smaller, resulting in lower rates of contact with molecules in solution and ultimately leading to potentially lower reaction rates, as well. However, this was not observed, and we speculate that this change in grain size with increasing temperature was too small to have a significant contribution to the photocatalytic activity. Furthermore, there is no observed change in grain size with increasing annealing temperature past 600 °C, while significant reductions in photocatalytic activity are still observed, suggesting that optical properties dominate the contribution to activity in this case.

5.3. Photocatalytic Activity and Film Thickness

Figure 5a shows the half-order concentration curves for films annealed at 300 °C and having thickness varying from 200 to 600 nm. Again, a control measurement was made, which consisted of a sapphire substrate without a ZnO film incubated in an 8-ppm rhoB solution while irradiated with UV light (hollow squares, Figure 5a). We subtracted the control curve from measurements made in ZnO-incubated solution (filled triangles, diamonds, squares and circles in Figure 5a), with the resulting slopes of the best-fit representing the photocatalytic activities of the metal-oxide films. The half-order photocatalytic activities are plotted as a function of film thickness in Figure 5b and are shown in Table 3. An approximately linear increase in photocatalytic activity is observed with increasing film thickness, as shown via the solid line in Figure 5b.

In the previous section, we investigated photocatalytic activity with film annealing temperature while keeping film thickness constant. For increasing annealing temperature, there was a significant decrease in photoemission efficiency, but little change in overall surface morphology, at least with respect to the observable height of ZnO grain protrusions, as seen in Figure 1d,e. In this section, we discuss the photocatalytic activity as a function of film thickness with constant annealing temperature. A comparison of the excitonic peaks for films annealed at 300°C at a thicknesses of 600 and 200 nm shows no significant difference in excitonic photoemission efficiency with thickness, as seen in Figure 2a,b, respectively. However, inspection of the AFM images shown in Figure 1b,d indicates a doubling in the observable ZnO grain protrusion height when going from 200 to 600 nm-thick films.

Figure 5. Half-order concentration curves for films annealed at 300 °C and having thicknesses varying from 200 to 600 nm. Data is fit to the half-order integrated rate law, with the curves of best-fit shown as the solid lines.

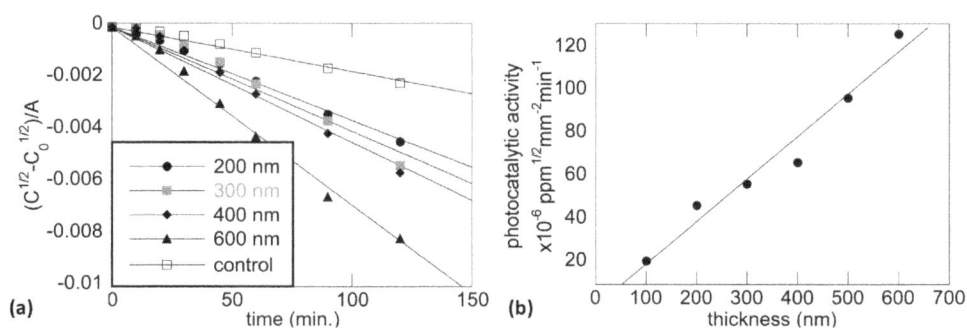

The significant increase in protrusion height could explain the increase in photocatalytic activity with increasing film thickness. As mentioned, the effective surface area on the nano-scale may also affect the photocatalytic activities of the films, though the precise extent of its significance has been under question [43]. The polycrystalline nature of our films results in approximately 100–150 nm-diameter ZnO columns that protrude upward from the surface, with an inter-column spacing on the order of 10–20 nm. Increasing the height of these columns results in an effective increase in surface area, resulting in higher rates of contact with molecules in solution and ultimately leading to higher reaction rates. The observed increase in green and yellow band emission with thicker films appears to have no effect on the photocatalytic activity of the films. Therefore, as expected, the excitonic photoemission efficiency and effective surface area of the catalytic material are the main contributors to activity. For films grown via a two-step thermal annealing process, low temperature annealing combined with relatively thick films results in materials demonstrating greater photocatalytic activity, due mainly to high optical quality and the porous nature of the morphology.

Table 3. Photocatalytic activity at various film thicknesses.

Film thickness (nm)	Photocatalytic activity ($ppm^{1/2}mm^{-2}min^{-1}$)
100	$(21 \pm 1) \times 10^{-6}$
200	$(47 \pm 1) \times 10^{-6}$
300	$(57 \pm 1) \times 10^{-6}$
400	$(67 \pm 1) \times 10^{-6}$
500	$(97 \pm 1) \times 10^{-6}$
600	$(127 \pm 1) \times 10^{-6}$

5.4. Film Stability

To determine film stability, we looked at the evolution of the photocatalytic activity over long periods of time (up to two days of continuous incubation). The order of the overall reaction can change over the course of the photocatalysis experiment due to decreasing concentration of rhoB or increasing concentrations of free radicals and peroxides in solution, which could manifest as a decrease in activity. We are interested in the photocatalytic activity of the films, so activity was determined via rhoB concentration measurements over one-hour intervals. The rhoB solution was replaced after each interval to ensure sufficient concentrations of rhoB and relatively low concentrations of free radicals and peroxides.

Figure 6 shows the photocatalytic activity as a function of time for a ZnO film having a thickness of 600 nm and annealed at a temperature of 300 °C. The photocatalytic activity decreases approximately linearly over 50 hours until reducing to zero. Interestingly, after 50 hours of continued incubation, the morphology of the sample is consistent with the morphology of the sapphire substrate (not shown), suggesting that the cause of the observed decreasing activity is the degradation of the ZnO film. During the incubation period, ZnO at the interface dissociates, forming Zn^+ and O^- ions that leave the film structure and enter the solution. This is consistent with reports in the literature concerning the stability of bulk and thin-film ZnO in contact with aqueous solutions [4,17]. Preferential etching should occur at termination sites, such as at the tips of ZnO grain protrusions seen in Figure 1b. This results in a slow decrease in film thickness as the reactions take place. As shown in the previous section, when the photoemission efficiency is similar, the predominant contributor to the photocatalytic activity is the specific surface area and, in the context of this study, the film thickness.

Therefore, the photocatalytic activity decreases with time due to surface etching of the film, resulting in decreasing film thickness.

Figure 6. Photocatalytic activity as a function of time for a ZnO film having a thickness of 600 nm and annealed at a temperature of 300 °C. Measurements where made spectrophotometrically over one-hour increments, with the rhoB solution being replaced after each one-hour measurement.

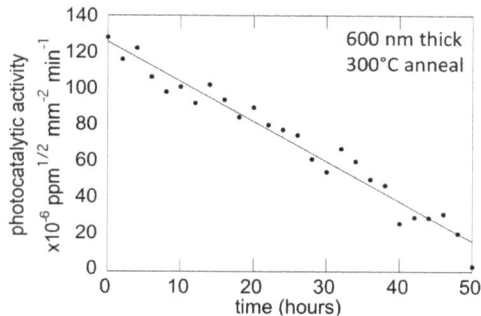

6. Discussion

Extensive work has gone into investigating the photocatalytic properties of ZnO nanoparticles and colloid suspensions [20]. However, separating the catalyst from solution is challenging, which makes their use in these applications potentially cost-prohibitive [22]. As mentioned, Zn toxicity is also a concern for these systems, especially if allowed to remain in the environment. Fujishima *et al.* suggest that the nanofilm form of these semiconductors is preferable to particles for use in fluid decontamination exactly because the nanoparticles need to eventually be collected and removed from the fluid [23,24]. There has been little discussion in the literature concerning the sorts of structural and photophysical properties that can directly affect photocatalytic activity for film-based catalysts. For the porous polycrystalline films discussed in this study, both optical and morphological properties have been shown to significantly affect the photodegradation of organics at the interface.

Polycrystalline films demonstrating high surface roughness and with small crystal grain size do typically have a high effective surface area, which increases the number of available surface states to serve as reaction sites. The concern with respect to the engineering of effective metal-oxide-based catalysts, though, is that the decreased crystallinity associated with these surfaces typically results in reduced optical efficiency and, therefore, lower electron-hole pair production efficiency [23,27,28]. Both high surface area and optical efficiency are required for high photocatalytic activity with metal-oxides. For TiO_2 and ZnO thin, polycrystalline films, a decrease in grain size typically results in increased surface roughness and surface area; however, small grain size typically also corresponds to an increase in deep-level defects that could lower the number of photo-induced holes at the surface available for catalysis [28]. These competing mechanisms must be balanced. Thermal oxidation of Zn-metal films at low annealing temperatures applied over many hours results in a balance between high effective surface area and optical quality. Furthermore, the porous nature of these films results in increased effective surface area with increased film thickness, without a corresponding decrease in optical quality, at least with respect to electron-hole pair production efficiency.

However, a significant problem with metal-oxide systems, such as ZnO, is long-term stability. The most photoactive films discussed in this study decrease in effectiveness by 50% in approximately 24 hours. There is also some concern about the dissolution of ZnO particles and resulting Zn toxicity in marine environments [4,17]. Although most of the work on ZnO dissolution has been with respect to nanoparticles and colloid suspensions, there is increasing evidence that Zn toxicity should be a concern for film-based systems and even for the interface of the single-crystal bulk [4,25]. Continued work needs to be done to find metal-oxide systems that exhibit high catalytic activity and that are passivated from preferential etching at termination sites, resulting in increased stability.

7. Conclusions

In summary, the photocatalytic activity and stability of thin, polycrystalline ZnO films was studied. The oxidative degradation of organic compounds at the ZnO surface results from the ultraviolet (UV) photo-induced creation of highly oxidizing holes and reducing electrons, which combine with surface water to form hydroxyl radicals and reactive oxygen species. Therefore, the efficiency of electron-hole pair formation is of critical importance for self-cleaning and antimicrobial applications with these metal-oxide catalyst systems. In this study, the lower annealing temperature of the fabricated ZnO thin films resulted in decreased protrusion size and specific surface area, as well as increased UV excitonic emission. The films oxidized at lower annealing temperatures exhibited higher photocatalytic activity, which is attributed to the increased optical quality. Photocatalytic activity was also found to depend on film thickness, with lower activity observed for thinner films due to a decrease in effective surface area. Decreasing activity with use was found to be the result of decreasing film thickness due to preferential surface etching.

Acknowledgments

This work was supported by the U.S. National Science Foundation Division of Materials Research (NSF-DMR 1104600) and the II-VI Foundation's Undergraduate Research Program. The photoluminescence data presented in this paper were obtained at Virginia Commonwealth University (Richmond, VA, USA) in the laboratory of Mikhail Reshchikov.

Author Contributions

This work was completed in Christopher Moore's research laboratory under his direction. Data collection and analysis was completed by all three authors. Robert Louder contributed writing to Sections 4 and 5. Cody Thompson contributed writing to Section 5. Both Robert Louder and Cody Thompson provided their main contributions to this work while undergraduate students at Coastal Carolina University. Robert Louder is currently a biophysics graduate student at the University of

California, Berkeley. Cody Thompson is currently a supervisor in the Research and Development laboratory at Wellman Engineering Resins.

Conflicts of Interest

The authors declare no conflict of interest.

References

1. Ozgur, U.; Alivov, Y.; Liu, C.; Teke, A.; Reshchikov, M.; Dogan, S.; Avrutin, V.; Cho, S.; Morkoc, H. A comprehensive review of ZnO materials and devices. *J. Appl. Phys.* **2005**, *98*, doi:10.1063/1.1992666.
2. Zhao, Q.; Xie, Y.; Zhang, Z.; Bai, X. Size-selective synthesis of zinc sulfide hierarchical structures and their photocatalytic activity. *Cryst. Growth Des.* **2007**, *7*, 153–158.
3. Diebold, U. Structure and properties of TiO$_2$ surfaces: A brief review. *Appl. Phys. A* **2003**, *76*, 681–687.
4. Moore, J.; Kenny, S.; Baird, C.; Morkoc, H.; Baski, A. Electronic behavior of the Zn- and O-polar ZnO surfaces studied using conductive atomic force microscopy. *J. Appl. Phys.* **2009**, *105*, doi:10.1063/1.3132799.
5. Covington, L.; Moore, J. Photoconductivity and transient response of Al:ZnO:Al planar structures fabricated via a thermal oxidation process. *Thin Solid Films* **2013**, *506*, 106–111.
6. Moore, J.; Thompson, C. A phenomenological model for the photocurrent transient relaxation observed in ZnO photodetector devices. *Sensors* **2013**, *13*, 9921–9940.
7. Ali, G.; Thompson, C.; Jasim, A.; Abdulbaki, I.; Moore, J. Effect of embedded Pd microstructures on the flat-band-voltage operation of room temperature ZnO-based propane gas sensors. *Sensors* **2013**, *13*, 16801–16815.
8. Ali, G.; Kadhim, A.; Thompson, C.; Moore, J. Electrical and optical effects of Pd microplates embedded in ZnO thin film based MSM UV photodetectors: A comparative study. *Sens. Actuators A* **2014**, *209*, 16–23.
9. Moore, J.; Covington, L.; Foster, R.; Gee, E.; Jones, M.; Morris, S. ZnO ultraviolet photodetectors grown via thermal oxidation of Zn-metal on glass and sapphire substrates. *Proc. SPIE* **2011**, *7940*, doi:10.1117/12.873837.
10. Gedamu, D.; Paulowicz, I.; Kaps, S.; Lupan, O.; Wille, S.; Haidarschin, G.; Mishra, Y.K.; Adelung, R. Rapid Fabrication Technique for Interpenetrated ZnO Nanotetrapod Networks for Fast UV Sensors. *Adv. Mater.* **2014**, *26*, 1541–1550.
11. Antoine, T.E.; Mishra, Y.K.; Trigilio, J.; Tiwari, V.; Adelung, R.; Shukla, D. Prophylactic, therapeutic and neutralizing effects of zinc oxide tetrapod structures against herpes simplex virus type-2 infection. *Antivir. Res.* **2010**, *96*, 363–375.
12. Hashimoto, K.; Irie, H.; Fujishima, A. TiO$_2$ Photocatalysis: A Historical Overview and Future Prospects. *Jpn. J. Appl. Phys.* **2005**, *44*, 8269–8285.

13. You, H.C.; Lin, Y.H. Investigation of the Sol-Gel Method on the Flexible ZnO Device. *Int. J. Electrochem. Sci.* **2012**, *7*, 9085–9094.

14. Moore, J.; Louder, R.; Covington, L.; Stansell, R. Photoconductivity and photocatalytic activity of ZnO thin films grown via thermal oxidation. *Proc. SPIE* **2012**, *8263*, doi:10.1117/12.906227.

15. Oppenlander, T. *Photochemical Purification of Water and Air*; Wiley-VCH: Weinheim, Baden-Württemberg, 2003.

16. Anpo, M. Utilization of TiO_2 photocatalysts in green chemistry. *Pure Appl. Chem.* **2000**, *72*, 1265–1270.

17. Bian, S.; Mudunkotuwa, I.; Rupasinghe, T.; Grassian, V. Aggregation and dissolution of 4 nm ZnO nanoparticles in aqueous environments: Influence of pH, ionic strength, size, and adsorption of humic acid. *Langmuir* **2011**, *27*, 6059–6068.

18. Juan, K. The fundamental and practice for the removal of phenolic compounds in wastewater. *Ind. Pollut. Prev. Control* **1984**, *3*, 88–92.

19. Carp, O.; Huisman, C.; Reller, A. Photoinduced reactivity of titanium dioxide. *Progr. Solid State Chem.* **2004**, *32*, 33–177.

20. Wan, Q.; Wang, T.; Zhao, J. Enhanced photocatalytic activity of ZnO nanotetrapods. *Appl. Phys. Lett.* **2005**, *87*, doi:10.1063/1.2034092.

21. Reimer, T.; Paulowicz, I.; Röder, R.; Kaps, S.; Lupan, O.; Chemnitz, S.; Benecke, W.; Ronning, C.; Adelung, R.; Mishra, Y.K. Single step integration of ZnO nano- and microneedles in Si trenches by novel flame transport approach: Whispering Gallery Modes and photocatalytic properties. *ACS Appl. Mater. Interfaces* **2014**, *6*, 7806–7815.

22. Konstantinou, I.; Albanis, T. TiO_2-assisted photocatalytic degradation of azo dyes in aqueous solution: Kinetic and mechanistic investigations. *Appl. Catal. B* **2004**, *49*, 1–14.

23. Fujishima, A.; Rao, T.; Tryk, D. Titanium dioxide photocatalysis. *J. Photochem. Photobiol. C Photochem. Rev.* **2000**, *1*, 1–21.

24. Chakrabarti, S.; Dutta, B. Photocatalytic degradation of model textiledyes in wastewater using ZnO as semiconductor catalyst. *J. Hazard. Mater. B* **2004**, *112*, 269–278.

25. Yonghao, W.; Feng, H.; Danmei, P.; Bin, L.; Dagui, C.; Wenwen, L.; Xueyuan, C.; Renfu, L.; Lin, Z. Ultraviolet-light-induced bactericidal mechanism on ZnO single crystals. *Chem. Commun.* **2009**, 6783–6785.

26. Fujishima, A.; Zhang, X.; Tryk, D. TiO_2 photocatalysis and related surface phenomena. *Surface Sci. Rep.* **2008**, *63*, 515–582.

27. Lin, D.; Wu, H.; Zhang, R.; Pan, W. Enhanced photocatalysis of electrospun Ag-ZnO heterostructured nanofibers. *Chem. Mater.* **2009**, *21*, 2479–3484.

28. Moore, J.; Covington, L.; Stansell, R. Affect of film thickness on the blue photoluminescence from ZnO. *Phys. Status Solidi A* **2012**, *209*, 741–745.

29. Sakthivel, S.; Kisch, H. Daylight photocatalysis by carbon-modified titanium dioxide. *Angew. Chem. Int. Ed.* **2003**, *351*, 1378–1383.

30. Covington, L.; Stansell, R.; Moore, J. Emergence of blue emission with decreasing film thickness and grain size for ZnO grown via thermal oxidation of Zn-metal films. *Mater. Res. Soc. Proc.* **2012**, *1394*, doi:10.1557/opl.2012.245.

31. Gupta, R.; Shridhar, N.; Katiyar, M. Structure of ZnO films prepared by oxidation of metallic zinc. *Mater. Sci. Semi. Proc.* **2002**, *5*, 11–15.

32. Chen, S.; Liu, Y.; Ma, J.; Zhao, D.; Zhi, Z.; Lu, Y.; Zhang, J.; Shen, D.; Fan, X. High-quality ZnO thin films prepared by two-step thermal oxidation of metallic Zn. *J. Cryst. Growth* **2002**, *240*, 467–472.

33. Cho, S.; Ma, J.; Kim, Y.; Sun, Y.; Wong, G.; Ketterson, J. Photoluminescence and ultraviolet lasing of polycrystalline ZnO thin films prepared by the oxidation of the metallic Zn. *Appl. Phys. Lett.* **1999**, *75*, doi:10.1063/1.125141.

34. Wang, Y.; Lau, S.; Lee, H.; Yu, S.; Tay, B.; Zhang, X.; Hng, H. Photoluminescence study of ZnO films prepared by thermal oxidation of Zn metallic films in air. *J. Appl. Phys.* **2003**, *94*, doi:10.1063/1.1577819.

35. Jain, A.; Sagar, P.; Mehra, R.M. Changes of structural, optical and electrical properties of sol-gel derived ZnO films with their thickness. *Mater. Sci. Pol.* **2007**, *25*, 233–242.

36. Wu, X.; Siu, G.; Fu, C.; Ong, H. Photoluminescence and cathodoluminescence studies of stoichiometric and oxygen-deficient ZnO films. *Appl. Phys. Lett.* **2001**, *78*, doi:10.1063/1.1361288.

37. Van Dijken, A.; Meulenkamp, E.; Vanmaekelbergh, D.; Meijerink, A. Identification of the transition responsible for the visible emission in ZnO using quantum size effects. *J. Lumin.* **2000**, *90*, 123–128.

38. Aida, M.; Tomasella, E.; Cellier, J.; Jacquet, M.; Bouhssira, N.; Abed, S.; Mosbah, A. Annealing and oxidation mechanism of evaporated zinc thin films from zinc oxide powder. *Thin Solid Films* **2006**, *515*, 1494–1499.

39. Zhao, J.; Hu, L.; Wang, Z.; Zhao, Y.; Liang, X.; Wang, M. High-quality ZnO thin films prepared by low temperature oxidation of metallic Zn. *Appl. Surf. Sci.* **2004**, *229*, 311–315.

40. Zhao, L.; Lian, J.S.; Liu, Y.H.; Jiang, Q. Influence of preparation methods on photoluminescence properties of ZnO films on quartz glass. *Trans. Nonferrous Metals Soc. China* **2008**, *18*, 145–149.

41. Fujishima, A.; Honda, K. Electrochemical photolysis of water at a semiconductor electrode. *Nature* **1972**, *238*, 37–38.

42. Fujishima, A.; Kobayakawa, K.; Honda, K. Hydrogen production under sunlight with an electrochemical photocell. *J. Electrochem. Soc.* **1975**, *122*, 1487–1489.

43. Wu,X.; Jiang, Z.; Liu, H.; Xin, S.; Hu, X. Photo-catalytic activity of titanium dioxide thin films prepared by micro-plasma oxidation method. *Thin Solid Films* **2003**, *441*, 130–134.

Tuning the Photocatalytic Activity of Anatase TiO$_2$ Thin Films by Modifying the Preferred <001> Grain Orientation with Reactive DC Magnetron Sputtering

Bozhidar Stefanov and Lars Österlund

Abstract: Anatase TiO$_2$ thin films were deposited by DC reactive magnetron sputtering on glass substrates at 20 mTorr pressure in a flow of an Ar and O$_2$ gas mixture. The O$_2$ partial pressure (P_{O_2}) was varied from 0.65 mTorr to 1.3 mTorr to obtain two sets of films with different stoichiometry. The structure and morphology of the films were characterized by secondary electron microscopy, atomic force microscopy, and grazing-angle X-ray diffraction complemented by Rietveld refinement. The as-deposited films were amorphous. Post-annealing in air for 1 h at 500 °C resulted in polycrystalline anatase film structures with mean grain size of 24.2 nm (P_{O_2} = 0.65 mTorr) and 22.1 nm (P_{O_2} = 1.3 mTorr), respectively. The films sputtered at higher O$_2$ pressure showed a preferential orientation in the <001> direction, which was associated with particle surfaces exposing highly reactive {001} facets. Films sputtered at lower O$_2$ pressure exhibited no, or very little, preferential grain orientation, and were associated with random distribution of particles exposing mainly the thermodynamically favorable {101} surfaces. Photocatalytic degradation measurements using methylene blue dye showed that <001> oriented films exhibited approximately 30% higher reactivity. The measured intensity dependence of the degradation rate revealed that the UV-independent rate constant was 64% higher for the <001> oriented film compared to randomly oriented films. The reaction order was also found to be higher for <001> films compared to randomly oriented films, suggesting that the <001> oriented film exposes more reactive surface sites.

Reprinted from *Coatings*. Cite as: Stefanov, B.; Österlund, L. Tuning the Photocatalytic Activity of Anatase TiO$_2$ Thin Films by Modifying the Preferred <001> Grain Orientation with Reactive DC Magnetron Sputtering. *Coatings* **2014**, *4*, 587-601.

1. Introduction

Photocatalytic and superhydrophillic thin films of anatase TiO$_2$ have attracted large research interest since the early 1970s and find application as self-cleaning, anti-fogging and anti-bacterial coatings [1–5]. Due to their intrinsic beneficial physical properties, TiO$_2$ films are suitable for architectural coatings, and commercially available self-cleaning glasses, tiles, cement, porcelain based on TiO$_2$ already exist [6–8].

There are a number of ways to deposit TiO$_2$ onto a glass substrate, including chemical methods such as dip-coating using sol-gel techniques [9], spray-pyrolysis [10], and atomic layer deposition [11]. Physical deposition techniques such as magnetron sputtering are industrial up-scalable technologies which allow for fast and controlled deposition of films with good optical and mechanical properties [12]. An added possibility for DC magnetron sputtering deposition of anatase TiO$_2$ is that the film textured can be controlled, and preferred orientated film growth in the <004> direction has been reported with improved photocatalytic properties [13].

Typically, the surfaces of TiO$_2$ nanocrystals are dominated by the stable {101} facets, with only a small amount of {001} facets, leading to a truncated bipyramidal shape [14,15]. The {001} facets have two times higher surface energy than {101} facets, but is expected to exhibit higher reactivity [16]. Ab initio calculations show that water dissociates on the (001) surface, while it only adsorbs molecularly on the (101) [17]. Altering the preparation conditions it was found that TiO$_2$ nanoparticles made by solvothermal synthesis can be prepared with altered <001>/<101> facet ratio [18–20]. In several studies it was reported that this led to an improved photocatalytic activity for the degradation of a number of liquid and gas-phase contaminants [21–23]. There are much fewer studies on sputter deposited TiO$_2$ films. It has been reported that sputtered TiO$_2$ films on biased substrates alter the texturing and preferred orientation depending on bias voltage, leading to an improved photodegradation rate of acetaldehyde as a function of increasing orientation [24]. Other reports show that changing total pressure [25], and O$_2$ partial pressure [26] may also affect film growth and preferential orientation of anatase films.

In the present study, we present results on the effect of preferred crystal grain orientation on the photocatalytic reactivity of anatase TiO$_2$ films prepared by DC magnetron sputtering. The preferred grain orientation is shown to be related to the partial pressure of O$_2$ in the reaction chamber. The effect of orientation on their photocatalytic properties was investigated. Both the apparent rate constant and intensity dependence of the photo-oxidation of methylene blue dye in liquid phase was studied. Our results show that even though the apparent reaction rate of the oriented films is higher, the increase in orientation does not affect the intensity dependence of the photo-degradation rate.

2. Experimental

2.1. Deposition of TiO$_2$ Anatase Thin Films

Anatase TiO$_2$ thin films were deposited by reactive DC magnetron sputtering on microscope slides glass substrates (Thermo Fischer Scientific, Waltham, MA, USA) using a Balzers UTT400 sputter system [27]. Two 5 cm diameter Ti targets (99.99% purity, Plasmaterials, Livermore, CA, USA) were used for deposition, and positioned 13 cm from the center of the sample holder, which was rotated at approximately 3 rpm, as described elsewhere [28]. Ar and O$_2$ gases (both with 99.997% purity) were supplied using mass flow controllers. The Ar flow rate was set to 60 mL·min^{-1} and the total pressure in the chamber was adjusted to 20 mTorr, and kept constant in the experiment, yielding approximately the same kinetic energy of the ions impinging on the substrate and thus facilitate inter-comparisons between fabrication conditions. The O$_2$ flow rate was varied in the experiments and two flow rates were employed, 2 mL·min^{-1} and 4 mL·min^{-1}, respectively, corresponding to an O$_2$ partial pressure of P_{O_2} = 0.65 mTorr and P_{O_2} = 1.3 mTorr in the sputtering chamber. One extra sample was deposited at intermediate O$_2$ pressure P_{O_2} = 0.95 mTorr (3 mL·min^{-1}) to confirm the dependence of physical properties on P_{O_2} (see Table 1), but this film was not studied further here. The plasma was created by a DC power supply set to constant current mode at 0.75 A yielding 212 W direct power at P_{O_2} = 0.65 mTorr O$_2$, and 245 W at P_{O_2} = 1.3 mTorr O$_2$.

At each O$_2$ partial pressures a total of 5 samples were sputtered. One sample with a size of 25 mm × 25 mm for X-ray diffraction measurements, and two batches consisting of two samples each deposited on larger substrates (50 mm × 25 mm, covered with a mask, thus limiting the coated area to 45 mm × 20 mm, or 900 mm^2) for the photo-catalytic experiments. Both sets of films were deposited for 35 min resulting in film thicknesses of 574 and 664 nm for the films sputtered at P_{O_2} = 0.65 and 1.3 mTorr O$_2$, respectively, as determined by surface profilometry (Bruker DektrakXT, Karlsruhe, Germany). The slight increase in sputtering rate from 16.4 to 19 nm·min^{-1} is likely to be due to the higher sputtering power obtained at 1.3 mTorr. The as-deposited samples were amorphous. To transform them into polycrystalline anatase they were calcined for 1 h at 500 °C. The temperature was ramped at 5 °C·min^{-1} and the samples were left to cool overnight.

2.2. Structural and Morphological Characterization

The film structure was investigated by grazing-incidence X-ray diffraction (GIXRD) employing a Siemens D5000 diffractometer equipped with parallel-beam optics and 0.4° Soller-slit collimator (Bruker AXS, Karlsruhe, Germany). The grazing angle was set to 0.5° and the diffractograms were collected with 5 s integration time and 0.05° resolution.

The film morphology was measured with scanning electron microscopy (SEM) using a FEI/Philips XL-30 environmental SEM microscope equipped with a field-emission gun (FEI, Hillsboro, OR, USA), and operated in Hi-Vac mode at 10 kV accelerating voltage.

The surface morphology was measured with atomic force microscopy (AFM) using a PSIA XE150 SPM/AFM (Park Systems Corp., Suwon, Korea) operating in non-contact mode in air at room temperature. Silicon ACTA cantilevers (AppNano, Mountain View, CA, USA) with 30 nm thick Al coating were used. The tip radius reported by the manufacturer was between 6 and 10 nm. Images were obtained at 1 Hz scanning rate over an area of 1000 nm × 1000 nm with a resolution of 256 pixels × 256 pixels.

2.3. Optical Measurements

Spectrophotometry using a Perkin-Elmer Lambda 900 spectrophotometer equipped with a 150 mm BaSO$_4$ coated integrating sphere was employed to optically determine the optical constant and film porosity. Transmission and reflectance spectra were recorded between 300 and 800 nm.

2.4. Photocatalytic Measurements

Photodegradation experiments of methylene blue (MB) dye in solution were used to quantify the photocatalytic activity of the two TiO$_2$ thin films with different preferred grain orientation. They will be referred to as sample "<101>" corresponding to the sample with 2% <001> orientation (i.e., almost randomly oriented grains dominated by <101> facets), and sample "<001>" with 25% <001> orientation. The experiments with MB were performed in a liquid-phase reactor employing in situ laser colorimetry for chemical analysis of the MB concentration employing a λ = 365 nm diode laser, as described in detail elsewhere [29]. A standard 4 W UV tube (λ = 365 nm) was used as a light source, positioned above the sample and attached on a stand, allowing for the distance between the tube and the reaction cell to be changed, as well as the illumination intensity. The intensity of the UV tube measured with a calibrated thermopile detector (Ophir, North Andover,

MA, USA) was 2.44 mW cm^{-2} at the UV tube's wall. Using the linear-source spherical emission (LSSE) model, the intensity distributions at the position of the film in the reactor was estimated to be 0.38 mW cm^{-2}.

Samples were placed on a sample holder at the bottom of the reaction cell, which was filled with 100 mL of distilled water and circulated with a magnetic stirrer. Background absorption in the reactor was measured, and then 1 mL of 100 ppm MB stock solution was added, leading to an initial concentration of 0.99 ppm. The system was allowed to reach equilibrium during a time period of 40 min allowing adsorption-desorption equilibrium between reactor walls and sample to be obtained. The MB concentration was measured in situ every 2 min during this equilibration and the increase of the laser signal at $\lambda = 365$ nm (due to MB adsorption in the reactor) was used to determine that equilibration was reached.

3. Results and Discussions

3.1. Film Structure and Morphology

It is evident from the diffractograms shown in Figure 1a that the as deposited films are completely amorphous. In contrast the heat-treated films consist only of anatase phase (Figure 1b). The different ratio between the <101> and <004> peaks in the two samples (<004> is here referred to as the <001> direction using conventional Miller indexing) suggests that they are textured and have different preferential orientation. To gain further insight, the diffractograms were Rietveld refined [30] using the PowderCell package [31] and compared with anatase crystallographic files [32]. The preferential orientation was approximated with the March-Dollase model [33], as implemented in PowderCell. This model is appropriate for films sputtered under rotation because it implies a cylindrical texturing symmetry. Rietveld refinement showed that the films consisted of polycrystalline anatase grains with mean crystallite size of 24.2 nm at $P_{O_2} = 0.65$ mTorr, and 22.1 nm at $P_{O_2} = 1.3$ mTorr. The <004> March-Dollase (MD) parameter was found to be 0.966 in the former case ($R_p = 5.9$, $R_{exp} = 3.49$) and 0.637 in the latter ($R_p = 5.7$, $R_{exp} = 2.94$). Using the Zolotoyabko equation [34], the MD parameters were converted into percentages of preferred orientation, viz.

$$\eta_{\langle hkl \rangle} = \frac{\left(1 - r_{\langle hkl \rangle}\right)^3}{\left(1 - r_{\langle hkl \rangle}^3\right)} \cdot 100\% \tag{1}$$

where $\eta_{\langle hkl \rangle}$ is the degree of preferential orientation in the <hkl> direction in % and $r_{\langle hkl \rangle}$ is the MD parameter calculated for this direction. The results from this analysis showed 2% preferential <001> orientation for the film sputtered at $P_{O_2} = 0.65$ mTorr, while 25% preferential orientation was found for the film sputtered $P_{O_2} = 1.3$ mTorr. The 2% oriented film thus corresponds to almost randomly oriented grains, yielding a stronger <101> peak, due to the high relative abundance of these crystal planes in the equilibrium anatase structure. Since the measurements were done at a grazing angle of 0.5° only the topmost 165 nm of the films are penetrated (and the information depth is even less). Thus, we can safely assume that the structure of the surface structure is consistent with the results from the GIXRD measurements. Considering a mean crystallite size of approximately 20 nm this implies that only a thin layer corresponding to 8 "particle layers" are probed. Below we refer to the sample sputtered at P_{O_2} = 0.65 mTorr to the <101> sample, and the one sputtered at P_{O_2} =1.3 mTorr to the <001> sample.

Qualitative comparison of the Ti Kα and O Kα peaks ratio from EDX showed that the heat treated samples have the same stoichiometry. For each partial O_2 pressure the ratio between the Ti and O peak was approximately constant at 0.65, confirming that the calcination in air oxidizes the substoichiometric films to an equilibrium structure (Table 1). Corresponding data for as-deposited films showed varying results due to bleaching (indicating gradual re-oxidation) of the films in the course of EDX analysis.

The SEM images shown in Figure 2 show that the films are composed of densely packed spherically shaped particles. No significant difference was observed between the two sets of films prepared at different O_2 pressures. Cross section images were obtained at 30° tilting angle and showed a dense film structure with no evidence of columnar growth.

Figure 1. XRD diffractograms of post-annealed (**a**) non-oriented and (**b**) <001> oriented films (the intensity ordinates for the two samples are shifted and given in arbitrary units). Photographs of the corresponding as-deposited films are shown on as insets on the right.

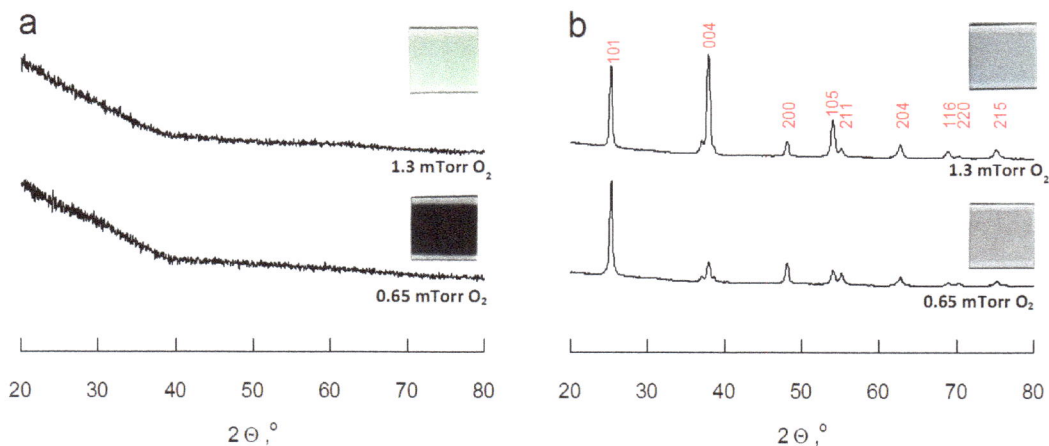

Figure 2. Scanning electron microscope micrographs showing top-view and cross sections (taken at 30° angle of incidence) of the post-annealed films sputtered at (**a,c**) P_{O_2} = 0.65 mTorr, and (**b,d**) P_{O_2} = 1.3 mTorr.

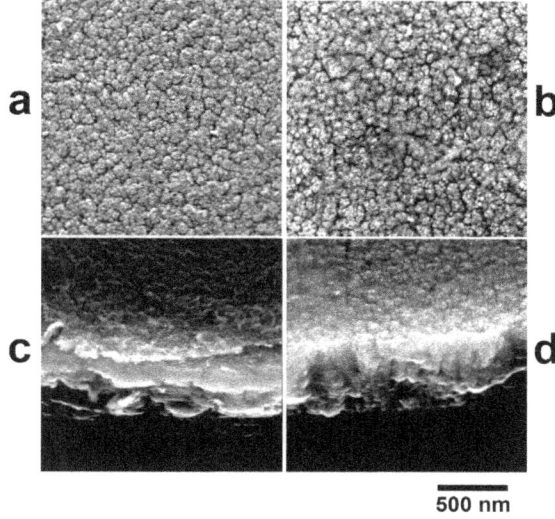

500 nm

It is apparent from the AFM images (Figure 3) that the surface morphology appears similar to the results from SEM. The raw data was treated to correct for Z-scanner error in the Y direction and exported as numerical values for statistical treatment in the R environment [35] using homemade scripts. Two surface morphology parameters were calculated: the root mean square (rms) surface roughness, R_q, and the average surface roughness wavelength, λ_q, where the latter is defined as

$$\lambda_q = 2\pi \frac{R_q}{\Delta_q} \tag{2}$$

where Δ_q is the rms surface slope, defined as

$$\Delta_q = \sqrt{\frac{1}{N-1} \sum_{N-1} \left(\frac{\Delta Z}{\Delta x}\right)^2} \tag{3}$$

where ΔZ is the change in height for every tip movement; Δx, in the x direction. The data for both R_q, Δ_q and λ_q were averaged for each of the 256 scan lines, then over the trace and retrace images, and then for three different measurements at random positions over the sample surface. Typical images for both films are shown in Figure 3. The rms surface roughness was estimated to be R_q = 1.31 ± 0.04 nm and 1.41 ± 0.04 nm, respectively, for the films sputtered at P_{O_2} = 0.65 and 1.3 mTorr. The Δ_q values were determined to be Δ_q = 0.105 ± 0.003 nm and 0.123 ± 0.004 nm, respectively, which yielded an average surface wavelength of λ_q = 77.9 ± 4.03 and 73.9 ± 3.72 for films prepared at P_{O_2} = 0.65 and 1.3 mTorr, respectively. Based on these results we can conclude that the films are smooth and are similar for the two sets of films, with slightly larger surface roughness and surface roughness wavelength for the preferentially <001> oriented films (within 8%). The physical properties of the anatase TiO_2 thin films are compiled in Table 1.

Figure 3. Atomic force microscopy (AFM) micrographs for the films with (**a**) 2% and (**b**) 25% <001> orientation.

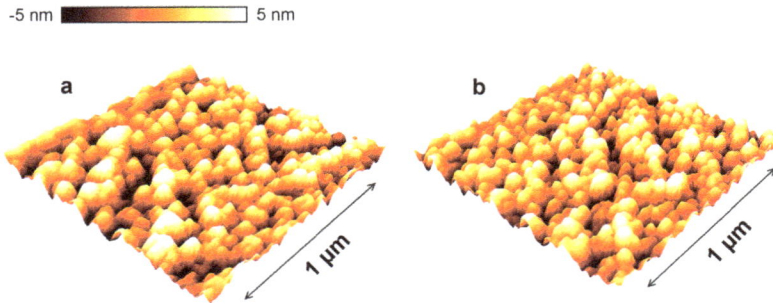

Table 1. Thin film deposition parameters and their effect on the physical properties of three samples deposited at similar conditions, but at different partial O_2 pressure, P_{O_2}.

Deposition parameters and physical properties	$P_{O_2} = 0.65$ mTorr	$P_{O_2} = 0.95$ mTorr	$P_{O_2} = 1.30$ mTorr
Total pressure (mTorr)		20	
Total flow (mL·min^{-1})	62	63	64
O_2 flow (mL·min^{-1})	2	3	4
Ar flow, (mL·min^{-1})	60	60	60
Sputtering power (W)	212	224	245
Sputtering rate (nm·min^{-1})	16	17	19
Stoichiometry of post-annealed films (from EDX, expressed as $I_{O\,K\alpha}/I_{Ti\,K\alpha}$)	0.652	0.650	0.645
Morphology			
Thickness, d (nm)	574	579	664
AFM rms roughness (nm)	1.3	1.1	1.4
Crystallographic properties			
March-Dollase (MD) parameter	0.966	0.810	0.637
Preferential <004> orientation, %	2	12	25
Mean crystalline size (nm)	24.2	26	22.1
Optical properties			
Refractive index, n	2.23	2.15	2.04
Packing density (Pullker), %	88	84	79
Optical bandgap, E_g (eV)	3.32	3.29	3.29

3.2. Optical Measurements

Spectrophotometry was employed to determine the optical constants and film porosity. The corresponding transmittances at 500 nm were measured to be 74% and 41%, respectively, for the as-deposited films, with the lower value for the films prepared at low P_{O_2}, which appear dark, almost black (Figure 1a, inset). After calcination the reflectance increased to about 70%–80% for all films, and they all became transparent with a slight visible tint due to light interference.

The optical constants and thicknesses of the two sets were determined using the envelope method suggested by Swanepoel [36]. The maxima and minima of the interference fringes at the transmittance spectrum were fitted with a set of spline functions, enveloping the spectrum, as depicted in Figure 4. The refractive index was then calculated using Equation (4).

$$n = \left[N + \left(N^2 - s^2 \right)^{\frac{1}{2}} \right]^{\frac{1}{2}}$$ (4)

where s is the refractive index of the substrate and N is defined as

$$N = 2s \frac{T_M - T_m}{T_M T_m} + \frac{s^2 + 1}{2}$$ (5)

where T_M and T_m are the maximum and minimum of the transmittance at a given wavelength. The refractive index of the substrate was calculated using

$$s = \frac{1}{T} + \left(\frac{1}{T} - 1 \right)^{\frac{1}{2}}$$ (6)

where T is the transmittance measured at a given wavelength. For our glass substrates s was determined to be $s = 1.39$. The refractive indices were determined from the averaged refractive index over all visible maxima and minima, expect the ones near the bandgap, where the transmittance starts to decrease, and the error becomes larger. The average refractive indices for films sputtered at $P_{O_2} = 0.65$ and $P_{O_2} = 1.3$ mTorr obtained in this manner were determined to be $n = 2.23$ and $n = 2.04$, respectively, and did not vary much over the wavelength region depicted in Figure 4. Hence, the reported values of n were determined as an average of the positions marked with a dashed line in Figure 4.

Figure 4. Transmittance spectra for a (**a**) non-oriented film, and (**b**) <001> oriented film. Positions of maxima and minima of T used in the analysis of the optical data, as well as average T over the visible wavelength region, are denoted by dashed lines.

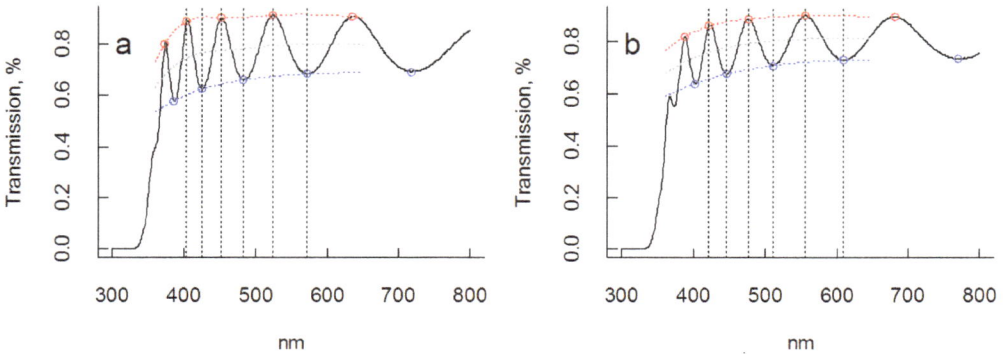

The thickness of the samples was determined using Equation (7):

$$d = \frac{\lambda_m \lambda_{m+1}}{2(\lambda_m \lambda_{m+1} - \lambda_{m+1} n_m)} \tag{7}$$

where λ_m, λ_{m+1} and n_m, n_{m+1} are the wavelength and the corresponding refractive index calculated by means of Equation (4) for any consecutive pair of maxima or minima in the UV-Vis transmittance spectra. The thicknesses calculated by means of Equation (7) were calculated to be $d = 591$ and 739 nm, respectively, for $P_{O_2} = 0.65$ and 1.3 mTorr, in good agreement with the results obtained from profilometry.

Changes in the refractive index are most likely to be associated with changes in sample porosity due different deposition conditions. The packing densities, ρ, of the two samples were therefore estimated using the Pulker equation [37].

$$\rho = \frac{\rho_f}{\rho_b} = \frac{n_f^2 - 1}{n_f^2 + 2} \cdot \frac{n_b^2 + 2}{n_b^2 - 1} \tag{8}$$

where ρ is the packing density of the sample; ρ_f and ρ_b the film and the bulk density of the material; and n_f and n_b the sample and bulk refractive index, respectively. The packing densities, corresponding to the measured refractive indices were determined to be 0.88 and 0.79, respectively, for the samples sputtered at $P_{O_2} = 0.65$ and 1.3 mTorr, respectively, which corresponds to a difference in porosity of about 10%, *i.e.*, a slightly increasing porosity with larger P_{O_2}, *i.e.*, a similar trend as for the surface roughness.

The optical bandgap, E_g, of the two films were calculated from the special absorption according to Hong *et al.* [38]

$$\alpha = \frac{1}{d} \ln \left(\frac{1-R}{T} \right) \tag{9}$$

where d is the thickness, T is the transmittance and, R the reflectance.

Since anatase TiO_2 is an indirect bandgap semiconductor, a plot of $\sqrt{\alpha E}$ as a function of photon energy, $E = h\nu$, should yield a linear dependence assuming parabolic band dispersion. This is a good approximation close to E_g ($E > E_g$). A linear region is discerned for both films in the region ~3.4 to ~3.6 eV, and extrapolation yields $E_g \approx 3.3$ eV for both samples in fair agreement with tabulated data of bulk anatase TiO_2 ($E_g = 3.2$ eV). Corresponding Tauc plots of $\sqrt{\alpha E}$ versus E are shown in Figure 5.

3.3. Photocatalytic Properties

Figure 6a,b show the results of the photocatalytic measurements. From the adsorption isotherms thus obtained the MB saturation coverage was determined, and found to be $8.17 \pm 0.6 \times 10^{-3}$ $\mu mol \cdot cm^{-2}$ for the <101> sample and $8.29 \pm 0.8 \times 10^{-3}$ $\mu mol \cdot cm^{-2}$ for the <001> sample, *i.e.*, an increase of 1.5% for the latter film. This is much smaller than the 10% increase in porosity inferred from analysis of the optical data described in Section 2.3, suggesting that the difference in porosity is related to pores inside the film structure which are inaccessible for the MB adsorption. Control experiments in a reactor containing uncoated glass substrates show that MB adsorption on reactor surfaces, other than TiO$_2$ film, is about 12% of the initial concentration.

Figure 5. Tauc plots of $\sqrt{\alpha E}$ *vs.* photon energy, E, and least square linear fit of data for films with (**a**) no orientation and (**b**) preferential <001> orientation. In both cases the optical bandgap was estimated to be 3.3 eV.

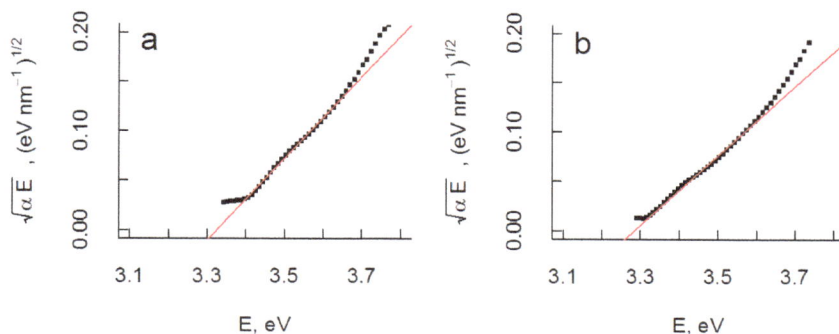

After MB equilibration the UV lamp was switched on and the photodegradation of MB was measured by the increasing colometric signal. The photodegradation of MB was modeled as a pseudo-first order reaction with the apparent rate k', viz.:

$$\frac{dC}{dt} = -k't \tag{10}$$

or

$$-\ln\frac{C}{C_0} = k't \tag{11}$$

where C_0 is the initial concentration of MB (after 40 min equilibration); C is the concentration at time t. Figure 6b shows a plot of Equation (11) for a <101> and a <001> film. The apparent rate constants averaged over a set of four samples from each batch were determined to be $k'_{<101>} = 0.99 \pm 0.09 \times 10^{-3}$ min^{-1} and $k'_{<001>} = 1.29 \pm 0.17 \times 10^{-3}$ min^{-1}, corresponding to an increase of k' by approximately 30% for the preferentially <001> oriented films compared to randomly oriented films.

The effect of UV intensity on the photocatalytic rate for the two sets of films was investigated. Repeated experiments were performed where the distance between the sample and the UV light source was systematically varied. It was changed from 8.5 cm, which is the closest distance, limited by the height of the reaction cell, up to 13.5 cm, 18.5 cm and 23.5 cm, yielding UV intensities at the sample position of 0.382 mW·cm^{-2}, 0.158 mW·cm^{-2}, 0.085 mW·cm^{-2} and 0.053 mW·cm^{-2}, respectively. The dependence of the rate constant of the UV light intensity was fitted using Equation (12):

$$k = k''I^\alpha \tag{12}$$

where k'' is the intensity independent rate constant, k is the apparent rate constant, I is the UV light intensity, and α is the reaction order by light intensity.

Figure 6. (a) Adsorption of methylene blue dye as a function of time, and **(b)** semi-logarithmic plot of the normalized MB concentration, C, as a function of UV irradiation time over <001> and <101> TiO$_2$ films. The gray line denotes the amount of dye adsorbed by reaction cell walls and uncoated glass slide in blank experiments.

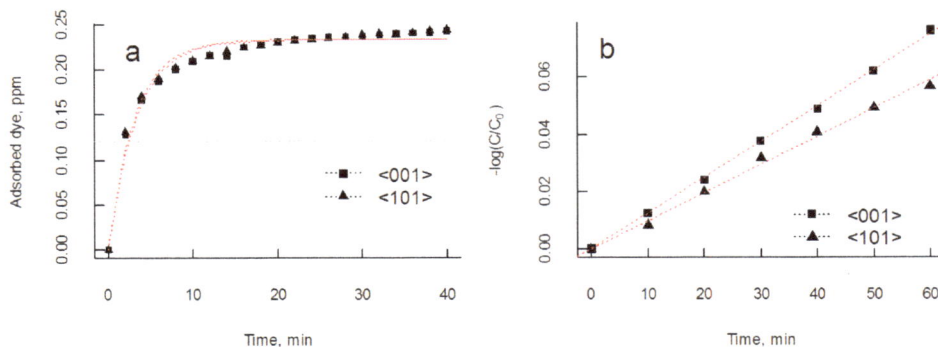

Figure 7 shows the effect of UV intensity on the photodegradation rate with different degree of preferential <001> orientation. In each case the experiments were conducted using four different samples. It was found that the increased orientation changes the way the catalyst is affected by the intensity of UV light. The less-oriented sample <101> showed an almost UV intensity independent rate constant for MB photodegradation with a rate constant $k''_{<101>} = 1.19 \times 10^{-3}$ min^{-1} and reaction order of $\alpha = 0.18$. In contrast the <001> sample yielded a rate constant of $k''_{<001>} = 1.95 \times 10^{-3}$ min^{-1} and a reaction order of $\alpha = 0.42$. Thus the UV independent rate constant k'' is 64% larger for the <001> oriented film. Again, this cannot be accounted for by a larger exposed surface area (higher porosity of surface roughness) as shown by the microscopy data, the estimates of film porosity, and surface coverage of MB. Instead it must be attributed to an intrinsic higher reactivity for the preferentially <001> oriented films. Mills and coworkers have reported that for thick, porous catalysts the rate constant, the intensity dependence can be divided into three regions [39]. The first region, at very low intensities, where the rate increases linearly with light intensity; a second region, at medium light intensities, where a square root dependence is observed, and a third region, at high intensities, where photon flux no longer limits the photo-degradation rate, and rate becomes independent of further increase of the light intensity. We can conclude from our measurements that for the randomly oriented grains dominated by {101} surfaces, the photo-degradation rate is not limited by UV intensity (with an almost constant rate as a function of UV intensity, $\alpha = 0.18$). Given that our films are thin and non-porous with undeveloped surface (based on the AFM measurements) we assign this to a small number of reactive sites and/or exposure of reactive sites with low reactivity. In contrast, for the preferentially <001> oriented films, we find $\alpha = 0.42$, which suggests that these films have a larger number of reactive sites and/or expose a larger fraction of more reactive sites (Table 2).

Figure 7. Methylene blue photo-degradation rate as a function of UV intensity for (**a**) a randomly oriented (<101>) TiO$_2$ film, and (**b**) a preferentially <001> orientated TiO$_2$ film. The dashed line is a guide to the eye.

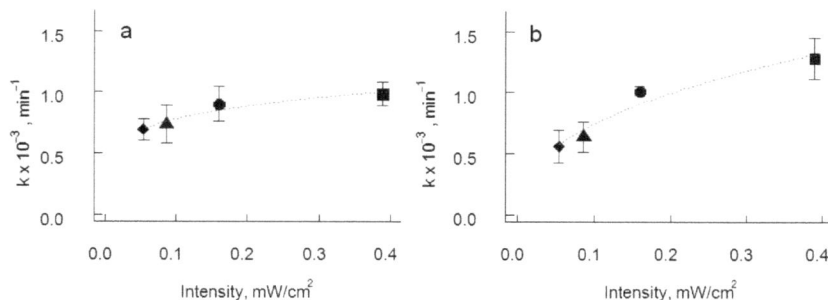

Furthermore, the UV intensity independent rate constant points to a total increase of activity by 61%. This is a dramatic increase, keeping in mind that the XRD analysis points to only 25% of the crystallites, oriented in the <001> direction (which does not directly translate to 25% increase of surface {001} coverage). For comparison, Yang *et al.* [40] demonstrated a novel synthesis of colloidal catalyst with 70% exposed highly-reactive {100} facets. Compared to Degussa P25 it showed 3-fold increase in the photo-oxidation rate of MB, which is similar to our results.

Table 2. Kinetic parameters for the photocatalytic degradation of MB for the samples with <001> preferential orientation, and with dominant <101> orientation (randomly oriented grains).

Samples	Orientation	
	<101> (random)	<001>
Apparent rate constant, $k'_{<hkl>}$ ($\times 10^{-3}$ min^{-1})	0.986 ± 0.09	1.286 ± 0.17
Intensity independent rate constant, $k''_{<hkl>}$ ($\times 10^{-3}$ min^{-1})	1.19	1.95
UV intensity reaction order, a	0.18	0.42

4. Conclusions

We have demonstrated that by purposefully adjusting the partial oxygen pressure in the sputtering chamber, reactive DC magnetron sputtering can be used to control the amount of preferential <001> orientation in TiO$_2$ thin films. The increased orientation was shown to lead to increased photocatalytic activity, which was attributed to exposure of a larger fraction of exposed reactive {001} facets at the film surface. The films with higher orientation also responded more strongly to changes in the intensity, yielding a significantly higher (64%) UV-independent photodegradation rate constant. Moreover, the reaction order was found to be almost independent of intensity for the <101> films ($\alpha = 0.18$), while it was $\alpha = 0.42$ for the preferentially oriented <001> films, suggesting that the preferentially <001> oriented films expose more reactive sites.

98

Acknowledgements

This work was funded by the European Research Council under the European Community's Seventh Framework Program (FP7/2007-2013)/ERC Grant Agreement No. 267234 ("GRINDOOR").

Author Contributions

Bozhidar Stefanov carried out all experiments, performed the data analysis and wrote the original manuscript. Lars Österlund conceived the idea, designed the structure of article and revised the manuscript.

Conflicts of Interest

The authors declare no conflict of interest.

References

1. Paz, Y.; Luo, Z.; Rabenberg, L.; Heller, A.J. Photooxidative self-cleaning transparent titanium-dioxide films on glass. *Mater. Research. Soc.* **1995**, *10*, 2842–2848.
2. Mellott, N.P.; Durucan, C.; Pantano, C.G.; Guglielmi, M. Commercial and laboratory prepared titanium dioxide thin films for self-cleaning glasses: Photocatalytic performance and chemical durability. *Thin Solid Films* **2006**, *502*, 112–120.
3. Ganesh, V.A.; Raut, H.K.; Nair, A.S.; Ramakrishna, S. A review on self-cleaning coatings. *J. Mater. Chem.* **2011**, *21*, 16304–16322.
4. Evans, P.; Sheel, D.W. Photoactive and antibacterial TiO_2 thin films on stainless steel. *Surf. Coat. Technol.* **2007**, *201*, 9319–9324.
5. Fu, G.; Vary, P.S.; Lin, C.-T. Anatase TiO_2 nanocomposites for antimicrobial coatings. *J. Phys. Chem. B.* **2005**, 109, 8889–8898.
6. Parkin, I.P.; Palgrave, R.G. Self-cleaning coatings. *J. Mater. Chem.* **2005**, *15*, 1689–1695.
7. Mills, A.; Lepre, A.; Elliott, N.; Bhopal, S.; Parkin, I.P.; O'Neill, S.A. Characterisation of the photocatalyst Pilkington Activ (TM): A reference film photocatalyst? *J. Photochem. Photobiol. A.* **2003**, *160*, 213–224.
8. Guo, S.; Wu, Z.-B.; Zhao, W.-R. TiO_2-based building materials: Above and beyond traditional applications. *Chin. Sci. Bull.* **2009**, *54*, 1137–1142.
9. Matsuda, A.; Kotani, Y.; Kogure, T.; Tatsumisago, M.; Minami, T. Transparent anatase nanocomposite films by the sol-gel process at low temperatures. *J. Am. Ceram. Soc.* **2000**, *83*, 229–239.
10. Abou-Helal, M.O.; Seeber, W.T. Preparation of TiO_2 thin films by spray pyrolysis to be used as a photocatalyst. *App. Surf. Sci.* **2002**, *195*, 53–62.
11. Liu, G.X.; Shan, F.K.; Lee, W.J.; Lee, G.H.; Kim, I.S.; Shin, B.C.; Yoon, S.G.; Cho, C.R. Transparent titanium dioxide thin film deposited by plasma-enhanced atomic layer deposition. *Integrat. Ferroelectr.* **2006**, *81*, 239–248.

12. Haseeb, A.S.M.A.; Hasan, M.M.; Masjuki, H.H. Structural and mechanical properties of nanostructured TiO$_2$ thin films deposited by RF sputtering. *Surf. Coat. Technol.* **2010**, *205*, 338–344.

13. Howitt, D.G.; Harker, A.B. The oriented growth of anatase in thin films of amorphous titania. *J. Mater. Res.* **1987**, *2*, 201–210.

14. Diebold, U.; Ruzycki, N.; Herman, G.; Selloni, A. One step towards bridging the materials gap: Surface studies of TiO$_2$ anatase. *Catal. Today* **2003**, *85*, 93–100.

15. Lazzeri, M.; Vittadini, A.; Selloni, A. Structure and energetics of stoichiometric TiO$_2$ anatase surfaces. *Phys. Rev. B.* **2001**, *63*, doi:10.1103/PhysRevB.63.155409.

16. Gong, X.; Selloni, A. Reactivity of anatase TiO$_2$ nanoparticles: The role of the minority (001) surface. *J. Phys. Chem. B* **2005**, *109*, 19560–19562.

17. Vittadini, A.; Selloni, A.; Rotzinger, F.P.; Grätzel, M. Structure and energetics of water adsorbed at TiO$_2$ anatase (101) and (001) surfaces. *Phys. Rev. Lett.* **1998**, *81*, 2954–2957.

18. Nguyen, C.K.; Cha, H.G.; Kang, Y.S. Axis-Oriented, Anatase TiO$_2$ Single Crystals with Dominant {001} and {100} Facets. *Cryst. Growth Des.* **2011**, *11*, 3947–3953.

19. Liu, B.; Aydil, E.S. Anatase TiO$_2$ films with reactive {001} facets on transparent conductive substrate. *Chem. Commun.* **2011**, *47*, 9507–9509.

20. Wang, L.; Zang, L.; Zhao, J.; Wang, C. Green synthesis of shape-defined anatase TiO$_2$ nanocrystals wholly exposed with {001} and {100} facets. *Chem. Commun.* **2012**, *48*, 11736–11738.

21. Zhang, J.; Chen, W.; Xi, J.; Ji, Z. {001} Facets of anatase TiO$_2$ show high photocatalytic selectivity. *Mater. Lett.* **2012**, *79*, 259–262.

22. Cao, F.-L.; Wang, J.-G.; Lv, F.-J.; Zhang, D.-Q.; Huo, Y.-N.; Li, G.-S.; Li, H.-X.; Zhu, J. Photocatalytic oxidation of toluene to benzaldehyde over anatase TiO$_2$ hollow spheres with exposed {001} facets. *Catal. Commun.* **2011**, *12*, 946–950.

23. Li, B.; Zhao, Z.; Gao, F.; Wang, X.; Qiu, J. Mesoporous microspheres composed of carbon-coated TiO$_2$ nanocrystals with exposed {001} facets for improved visible light photocatalytic activity. *App. Catal. B.* **2014**, *147*, 958–964.

24. Lyandres, O.; Finkelstein-Shapiro, D.; Chakthranont, P.M.; Graham, K.; Gray, A. Preferred Orientation in Sputtered TiO$_2$ Thin Films and Its Effect on the Photo-Oxidation of Acetaldehyde. *Chem. Mater.* **2012**, *24*, 3355–3362.

25. Meng, L.-J.; dos Santos, M.P. The influence of oxygen partial-pressure on the properties of dc reactive magnetron-sputtered titanium-oxide films. *Thin Solid Films* **1993**, *226*, 22–29.

26. Sério, S.; Melo Jorge, M.E.; Maneira, M.J.P.; Nunes, Y. Influence of O$_2$ partial pressure on the growth of nanostructured anatase phase TiO$_2$ thin films prepared by DC reactive magnetron sputtering. *Mat. Chem. Phys.* **2011**, *126*, 73–81.

27. Le Bellac, D.; Niklasson, G.A.; Granqvist, C.G. Angular-selective optical transmittance of anisotropic inhomogeneous Cr-based films made by sputtering. *J. Appl. Phys.* **1995**, *77*, 6145–6151.

28. Kharrazi, M.; Azens, A.; Kullman, L.; Granqvist, C.G. High-rate dual-target dc magnetron sputter deposition of electrochromic MoO$_3$ films. *Thin Solid Films* **1997**, *295*, 117–121.

29. Stefanov, B.I.; Kaneva, N.V.; Puma, G.L.; Dushkin, C.D. Novel integrated reactor for evaluation of activity of supported photocatalytic thin films: Case of methylene blue degradation on TiO_2 and nickel modified TiO_2 under UV and visible light. *Colloids Surf. A* **2011**, *382*, 219–225.

30. Rietveld, H.M. A profile refinement method for nuclear and magnetic structures. *J. Appl. Crystallogr.* **1969**, *2*, 65–71.

31. Kraus, W.; Nolze, G. POWDER CELL—A program for the representation and manipulation of crystal structures and calculation of the resulting X-ray powder patterns. *J. Appl. Crystallogr.* **1996**, *29*, 301–303.

32. Horn, M.; Schwerdtfeger, C.F.; Meagher, E.P. Refinement of structure of anatase at several temperatures. *Z. Krist.* **1972**, *136*, 273–281.

33. Dollase, W.A. Correction of intensities for preferred orientation in powder diffractometry—Application of the march model. *J. Appl. Crystallogr.* **1986**, *19*, 267–272.

34. Zolotoyabko, E. Determination of the degree of preferred orientation within the March-Dollase approach. *J. Appl. Crystallogr.* **2009**, *42*, 513–518.

35. *R*—A language and environment for statistical computing; R Foundation for Statistical Computing: Vienna, Austria. Available online: http://www.R-project.org/ (accessed on 9 July 2014).

36. Swanepoel, R. Determination of the thickness and optical-constants of amorphous-silicon. *J. Phys. E Sci. Instrum.* **1983**, *16*, 1214–1222.

37. Pulker, H.K. Characterization of optical thin-films. *Appl. Optic.* **1979**, *18*, 1969–1977.

38. Hong, S.; Kim, E.; Kim, D.-W.; Sung, T.-H.; No, K. On measurement of optical band gap of chromium oxide films containing both amorphous and crystalline phases. *J. Non-Cryst. Solids* **1997**, *221*, 245–254.

39. Mills, A.; Wang, J.; Ollis, D.F. Kinetics of liquid phase semiconductor photoassisted reactions: Supporting observations for a pseudo-steady-state model. *J. Phys. Chem. B* **2006**, *110*, 14386–14390.

40. Yang, Y.; Wang, G.; Deng, Q.; Kang, S.; Ng, D.H.L.; Zhao, H. A facile synthesis of single crystal TiO_2 nanorods with reactive {100} facets and their enhanced photocatalytic activity. *CrystEngComm* **2014**, *16*, 3091–3096.

Photocatalytic TiO₂ and Doped TiO₂ Coatings to Improve the Hygiene of Surfaces Used in Food and Beverage Processing—A Study of the Physical and Chemical Resistance of the Coatings

Parnia Navabpour, Soheyla Ostovarpour, Carin Tattershall, Kevin Cooke, Peter Kelly, Joanna Verran, Kathryn Whitehead, Claire Hill, Mari Raulio and Outi Priha

Abstract: TiO₂ coatings deposited using reactive magnetron sputtering and spray coating methods, as well as Ag- and Mo-doped TiO₂ coatings were investigated as self-cleaning surfaces for beverage processing. The mechanical resistance and retention of the photocatalytic properties of the coatings were investigated over a three-month period in three separate breweries. TiO₂ coatings deposited using reactive magnetron sputtering showed better mechanical durability than the spray coated surfaces, whilst the spray-deposited coating showed enhanced retention of photocatalytic properties. The presence of Ag and Mo dopants improved the photocatalytic properties of TiO₂ as well as the retention of these properties. The spray-coated TiO₂ was the only coating which showed light-induced hydrophilicity, which was retained in the coatings surviving the process conditions.

Reprinted from *Coatings*. Cite as: Navabpour, P.; Ostovarpour, S.; Tattershall, C.; Cooke, K.; Kelly, P.; Verran, J.; Whitehead, K.; Hill, C.; Raulio, M.; Priha, O. Photocatalytic TiO₂ and Doped TiO₂ Coatings to Improve the Hygiene of Surfaces Used in Food and Beverage Processing—A Study of the Physical and Chemical Resistance of the Coatings. *Coatings* **2014**, *4*, 433-449.

1. Introduction

In aquatic environments, microorganisms have a tendency to attach to surfaces along with organic and inorganic soil. For example in breweries, microorganisms have been shown to accumulate on sterile stainless steel surfaces within hours after the start of production [1].

Consumer demand is driving the development of a new group of more sensitive beverages with less alcohol, hop substances, and preservatives; however, these products are more prone to spoilage than are traditional drinks [2]. There are numerous operations involved in making beer. Each stage has a level of cleanliness that needs to be achieved and fouling is encountered at each stage [3]. Attachment of primary colonizers to stainless steel has been shown to be increased by sugars and sweeteners [1]. Thus removal of these deposits is essential since conditioning of a surface may be followed by biofilm formation. Biofilms on bottling plant surfaces are considered as serious sources for potential product spoiling microorganisms in the brewing industry [4]. Further, Fornalik [5] noted that minor fouling organisms resistant to cleaning in place (CIP) may become more resistant with time. Rheological studies indicated that increasing the temperature of the deposit generated a more elastic deposit which may decrease cleanability [3]. Thus, regular daily cleaning is needed. The following media are usually used in the cleaning process in brewing industry: water and steam, peroxide and alcohol based disinfectants, alkaline and acidic detergents and organic solvents [6,7]. There are however numerous drivers for a revision of CIP operations including the need to minimise utility usage (energy and water) and production downtime, minimisation of waste and greenhouse gas (GHG) emissions, and the need for product safety and quality [3].

One way to reduce cleaning costs and to improve process hygiene could be to use self-cleaning and antimicrobial coatings which can prevent the attachment of microorganisms and soil, or facilitate their efficient removal in the cleaning process.

TiO_2 is a widely used semiconductor. It has many different applications in optics [8], the environment [9], photovoltaics and solar cells [10,11], self-cleaning [12,13] and antimicrobial coatings [14]. In the self-cleaning and antimicrobial applications, the intended mechanism of action is often photocatalytic; in which the action of light on the TiO_2 coating generates active species that may be detrimental to microbes. For these applications, thin TiO_2 films with submicron thicknesses are usually employed. Several studies have been carried out to investigate the effect of crystal structure on the photocatalytic performance of TiO_2. Whilst some studies have found a higher activity of the anatase form [15,16], others have reported the mixed phase anatase/rutile to show a better photocatalytic performance [17]. Comparative studies of single phase anatase and rutile TiO_2 have concluded that the photocatalytic activity is dependent on the reaction being studied and different kinetics and intermediaries may be produced in each case [18,19]. As the surfaces used in the food and beverage industries are exposed to adverse environments (contact with water and beverages, cleaning solutions, abrasive wear during cleaning), scratch and corrosion resistance play important roles in their mechanical durability and chemical stability. Hence, it is important to satisfy several requirements, including good adhesion to the substrate, the retention of high activity and resistance to chemicals.

The adhesion of any film to its substrate is one of the most important properties of a thin film. The level of adhesion depends on the force required to separate atoms or molecules at the interface between film and substrate. The adhesion of a film to the substrate is strongly dependent on the chemical nature, cleanliness, and microscopic topography of the substrate surface [20]. The presence of contaminants on the substrate surface may increase or decrease the adhesion depending on whether the adsorption energy is increased or decreased, respectively. Also the adhesion of a film can be improved by providing more nucleation sites on the substrate, for instance, by using a fine-grained substrate or a substrate pre-coated with suitable materials. Of the deposition processes available, magnetron sputtering has been shown to produce well adhered and uniform coatings over wide areas [11]. In this process, the adhesion of the film to the substrate can be improved by ion-cleaning of the substrate prior to the coating deposition as well as additional ion bombardment during coating deposition which improves adhesion by providing intermixing on an atomic scale [21].

It has been shown throughout the literature that the chemical and structural properties of the active film have a profound impact on the overall photocatalytic performance. Photocatalytic performance is influenced by film characteristics including; composition, bulk and surface structure and nanostructure, atomic to nanoscale roughness, hydroxyl concentration, and impurity concentration (e.g., Fe and Cr) [22–25].

The work described in this paper investigates the chemical and mechanical durability, wettability and the retention of photocatalytic activity of selected coatings after being placed in different brewery process environments, in this case bottle/can filling lines in three Finnish breweries.

2. Experimental Section

2.1. Preparation of Coated Surfaces

The substrate material for all coatings was stainless steel AISI 304 2B ($75 \times 25 \times 1.6$ mm^3). Coatings were produced using either closed field unbalanced magnetron sputtering (CFUBMS) [21] or by spray-coating with a TiO$_2$ sol. Table 1 shows the coatings produced.

Table 1. Preparation method of coated surfaces.

Code	Coating	Deposition Method
T1	TiO$_2$	Reactive magnetron sputtering
T2	TiO$_2$-Ag (low)	–
T3	TiO$_2$-Ag (high)	–
U1	TiO$_2$	Reactive magnetron sputtering + heat treatment
U2	TiO$_2$-Mo	–
MC	TiO$_2$	Spray-coated with TiO$_2$ sol

Coatings T1–T3 were deposited using reactive magnetron sputtering in a Teer Coatings UDP 450 coating system. One titanium target (99.5% purity) was used for the deposition of TiO$_2$. Argon (99.998% purity) was used as the working gas and oxygen (99.5% purity) as the reactive gas. The working pressure was 1 mbar. Ag (99.95% purity) was used as the dopant. Advanced Energy Pinnacle Plus pulsed DC power supplies were used to power the titanium magnetrons and bias the substrates. An Advanced Energy DC power supply was used to power the silver target. 10–30 substrates were ultrasonically cleaned in acetone prior to loading into the chamber in order to remove surface contaminants. The substrates were aligned on a flat plate parallel to the surface of the metal targets at a distance of 150 mm from the target plane. A high rotational speed of 10 rpm was applied to the substrates to ensure enhanced mixing of silver and titanium within the coatings rather than the preferential formation of multilayer coatings. The substrates were ion-cleaned for a period of 20 min prior to the coating deposition using a bias voltage of -400 V and a low current of 0.2–0.35 A on the targets. The coatings were deposited at a bias voltage of -40 V. A thin layer of Ti was initially deposited as the adhesion layer prior to the introduction of oxygen to the deposition chamber. The amount of oxygen was controlled using an optical emission monitor, using conditions known to produce stoichiometric TiO$_2$ [26]. A pulsed-DC power of 2.5 kW was used on the Ti target at frequency 50 kHz and a duty of 97.75% (in synchronous mode). A continuous DC power of 70 W in the case of T2 and 150 W in the case of T3 was applied to the Ag target to vary the dopant content. The deposition rate was 17–22 nm/min depending on Ag content and coatings with thickness of 0.8–1 μm were produced. No additional heating was used during the coating process and the temperature did not exceed 200 °C during the process.

U1 and U2 coatings were deposited using reactive magnetron sputtering in a Teer Coatings UDP 450 coating system as described above. Two opposing magnetrons were fitted with titanium targets and one with the Mo dopant metal target (99.5% purity). The magnetrons with the titanium targets were in the closed field configuration and driven in pulsed DC sputtering mode using a dual channel Advanced Energy Pinnacle Plus supply at a frequency of 100 kHz and a duty of 50% (in synchronous mode). The Mo metal target was driven in a continuous DC mode (Advanced Energy MDX). The Ti targets were operated at a constant time-averaged power of 1 kW and the dopant target was operated at 180 W. Stainless steel samples were mounted on a substrate holder, which was rotated between the magnetrons at 4 rpm during deposition. The target to substrate separation was 8 cm. The titanium and Mo targets were cleaned by pre-sputtering in a pure argon atmosphere for 10 min. Deposition times were adapted to obtain a film thickness of 0.8–1 μm (deposition rate was 7.5 nm/min). The sputtered films were post deposition annealed at 600 °C for 30 min. in air.

Coating MC was prepared by spray-coating with a proprietary water-based TiO_2 sol using the following method. This transparent, neutral sol contained 2% TiO_2 (as anatase). Degreased stainless steel coupons were fixed to aluminium panels (approximately 150×100 mm^2). The panels with attached coupons were accurately weighed. The TiO_2 sol (0.2–0.3 g) was sprayed onto the aluminium panel with the attached coupons in a slow, steady motion, sweeping the panel in horizontal stripes from top to bottom, using a Badger Airbrush 200-3 model spray kit (Badger Air-Brush Co., Franklin Park, IL, USA). After air-drying for at least 15 min, the spraying procedure was repeated until 0.8–1.0 g/m^2 of TiO_2 sol was delivered to the surface. After air-drying overnight, the aluminium panel with the attached stainless steel coupons was re-weighed to give an accurate measurement of the weight per area of the coating.

2.2. Wettability

Water contact angle measurement is a practical tool to determine the wettability of a surface. Contact angle values were measured using a Digidrop instrument. At least two drops were measured for each surface and the measurements averaged. Measurements were conducted after exposure in light, either SUNTEST CPS+ (xenon arc, filtered with special window glass, 550 W/m^2 across the irradiance range 320–800 nm) or UVA light (Philips blacklight, 10–12 W/m^2 across the irradiance range 350–400 nm), or after storage in the dark.

2.3. Adhesion of Coatings

The scratch and wear resistance of the coatings were assessed using a Teer ST3001 scratch–wear tester (Teer Coatings Ltd, Droitwich, UK) [27]. The coated surfaces were evaluated using a Rockwell diamond tip (radius 200 μm). A load rate of 100 N·min^{-1} and a constant sliding speed of 10.0 mm·min^{-1} were used with the load increasing from 10 to 40 N. The scratch tracks were examined using a Cambridge Stereoscan 200 scanning electron microscope (Cambridge Instruments, Cambridge, UK) in order to detect any flaking.

2.4. Photocatalytic Characterization of Coatings

The photocatalytic activities of the coatings were analyzed using the methylene blue (MB) degradation assay under UV and fluorescent light sources. In brief, MB solutions were made up to an initial concentration of 0.0105 mMol·L^{-1}. Photocatalytic surfaces were placed in 10 mL of the MB solution and irradiated at an integrated power flux of 40 W/m^2 with two 15 W UV lamps (365 nm wavelength). Tests were also carried out using two 15 W fluorescent tubes in place of the UV tubes to simulate typical lighting environments. The integrated power flux to the coatings with the fluorescent tubes was 64 W/m^2, of which the UV component (300–400 nm) was 13 W/m^2. A 10 cm distance between the light source and MB solution was used. Samples of the MB solution were taken before testing and at 1 hour intervals up to a total of 5–8 h. and analyzed using a UV-Vis spectrophotometer (Perkin Elmer, Waltham, MA, USA). Spectra were taken in the range of 650–668 nm and the height of the absorption peak in this region was monitored.

A graph of peak height absorbance against irradiation time, which has an exponential form was generated. An index of photocatalytic activity (Pa) was defined by comparing the degradation rate of the MB solution in contact with the coated surfaces to the rate for an irradiated MB solution with no coating present. The equation below was used to calculate the photocatalytic activity of each of the films. Two parameters were defined: Pa_{UV} for UV irradiation and Pa_{FL} for fluorescent light irradiation [28].

$$Pa = 1 - C_0 \left[\frac{e^{-mx}}{e^{-cx}} \right] \tag{1}$$

where C_0 = peak height at time = 0; $C_0 e^{-mx}$ = decay rate of methylene blue; $C_0 e^{-cx}$ = decay rate of methylene blue in contact with photocatalytic coating.

2.5. Process Tests

Coated stainless steel pieces were placed on process surfaces within three breweries for a period of three months. Figure 1 shows an example of samples in location. Details of the location of test pieces in each brewery are given in Table 2. There was no special provision of lighting for the photocatalytic coatings; the process test took place under the usual brewery conditions of lighting, with coupons receiving varying amounts of light depending on their position in each machine. Furthermore, all samples underwent the normal process conditions and cleaning regimes used in each brewery which included acid and alkaline cleaning chemicals such as acetic acid and sodium hydroxide, ethanol, steam and mechanical brushing. For each coating, two replicates were used in each of the breweries. Additionally, two replicates were retained as controls and were kept in the dark for the same period. After three months, each replicate was cut into six sections. The mechanical durability, photocatalytic activity and wettability were evaluated each on two of these sections.

Figure 1. Samples in location at Brewery B.

Table 2. Coatings evaluated in process tests (for a period of three months).

Coating	Control	Brewery A [1]	Brewery B [2]	Brewery C [3]
TiO_2 (T1)	T1-R1 T1-R2	T1-1 T1-2	T1-3 T1-4	T1-5 T1-6
TiO_2-Ag (low) (T2)	T2-R1 T2-R2	T2-1 T2-2	T2-3 T2-4	T2-5 T2-6
TiO_2-Ag (high) (T3)	T3-R1 T3-R2	T3-1 T3-2	T3-3 T3-4	T3-5 T3-6
TiO_2 (U1)	U1-R1 U1-R2	U1-1 U1-2	U1-3 U1-4	U1-5 U1-6
TiO_2-Mo (U2)	U2-R1 U2-R2	U2-1 U2-2	U2-3 U2-4	U2-5 U2-6
TiO_2 (MC)	MC-R1 MC-R2	MC-1 MC-2	MC-3 MC-4	MC-5 MC-6

[1] Filler table of beer canning machine; [2] Seamer of beer canning machine; [3] Filler table of a water and soft drinks PET line, inclined 10°.

3. Results and Discussion

This work compared three TiO_2 surfaces: as-deposited and heat-treated coatings deposited by reactive magnetron sputtering (T1 and U1, respectively), and a spray-coated TiO_2 (MC). Two dopants (Ag and Mo) were also investigated. Ag was used as it is a well-known antimicrobial material which could impart additional antimicrobial functionality to the coating. Mo was used as a dopant to reduce the band gap of TiO_2 in order to improve the visible light activity of TiO_2. Mo-TiO_2 has been reported to shift the band gap of TiO_2 by -0.20 eV [28]. The photoactivity and mechanical properties of the surfaces were studied for the as-prepared coatings and those having undergone process conditions. The effect of the process conditions on the properties of the coatings was investigated.

3.1. As Prepared Coatings

SEM and EDX were used to analyze the topography and dopant concentration (as atomic percent of total metals) in the as-prepared doped coatings. Ag-TiO_2 and Mo-TiO_2 surfaces showed small submicron sized particles which were characterized by EDX as silver rich phases, suggesting that the dopant separated from the matrix TiO_2. The silver content was 0.50 ± 0.05 at% in T2 and 30.0 ± 3.1 at%

in T3. The Mo content in U2 was 7.0 ± 0.8 at %. The structure of coatings was analysed using XRD (Figure 2). The as-deposited TiO_2 coating (T1), showed an anatase structure. $Ag-TiO_2$ coatings showed strong silver peaks. The heat treated TiO_2 and $Mo-TiO_2$ (U1 and U2) showed anatase and rutile peaks as well as monoclinic β-TiO_2 which were very strong in the case of the doped coating.

Figure 2. Microstructure of coatings as evaluated using XRD, (**a**) as deposited TiO_2 and $Ag-TiO_2$ coatings (T1–T3); and (**b**) TiO_2 and $Mo-TiO_2$ coatings after heat treatment (U1 and U2) (S—substrate, An—anatase, Ru—rutile).

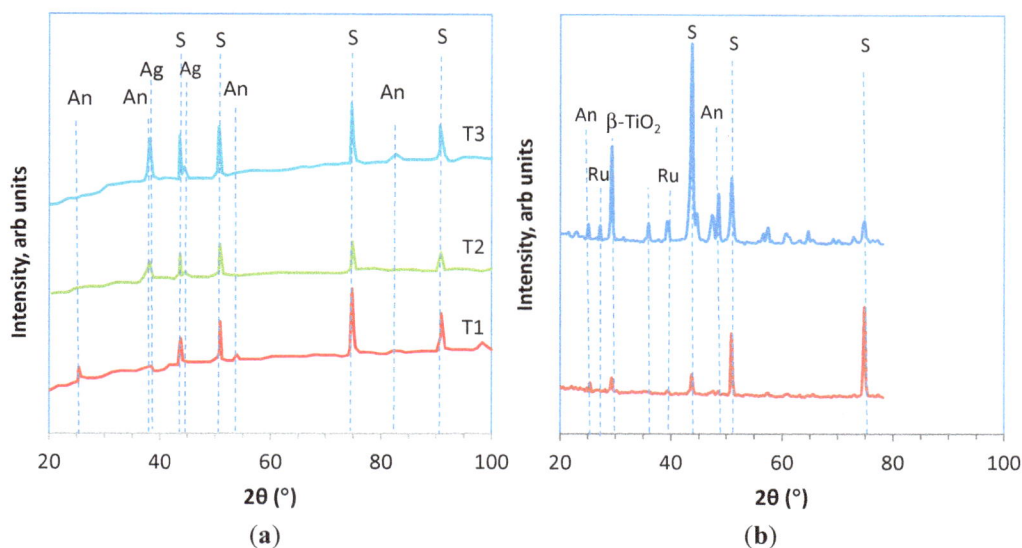

(**a**)   (**b**)

Figure 3 shows the photocatalytic activity for the as-prepared coatings and compares these values with those obtained for Pilkington Activ™ as a standard commercial product. As can be seen, all coatings showed high photocatalytic activity. In the case of T3, a change was also observed in the colour of the solution. This was thought to have been caused by leaching of silver from the surface. SEM analysis of the coating was performed before and after immersion in water for 2 h and showed the presence of microparticles on the surface which EDX confirmed to be silver (Figure 4). The silver microparticles in the as deposited coating were embedded in the matrix. Immersion in water resulted in the silver particles to protrude from the surface and EDX showed a reduction in the silver content, confirming that silver was indeed diffusing out of the coating.

Figure 3. Photocatalytic activity of the as-deposited coatings and comparison with a commercially available photocatalytic surface (Pilkington Activ™).

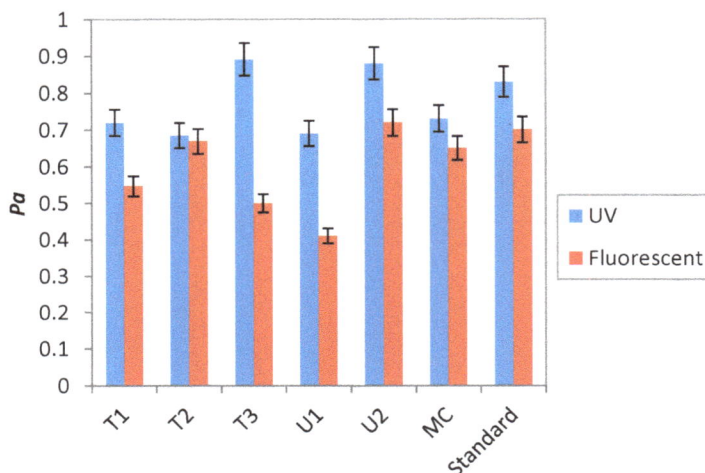

Figure 4. SEM micrographs of T3, (**a**) as deposited coating; and (**b**) after being under water for 2 h.

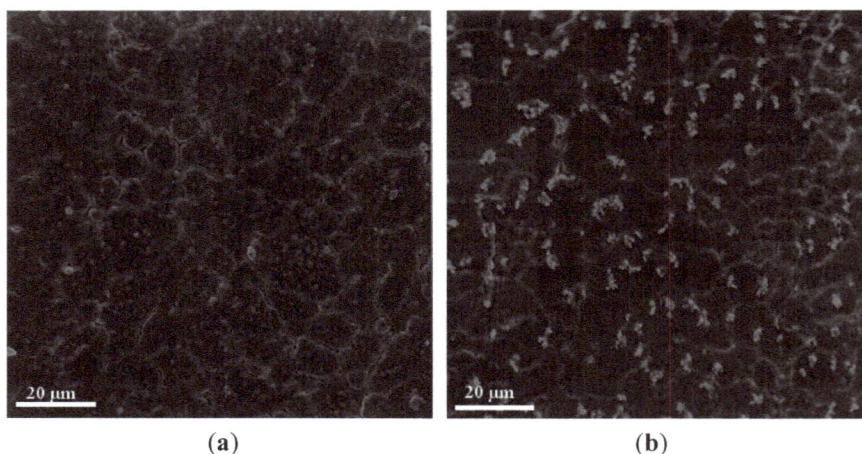

(**a**)   (**b**)

Mechanical resistance of the coatings was analyzed using scratch testing. Figure 5 shows the scratch tracks of the coatings after production, as observed using the SEM. Coatings T1–T3 and MC showed excellent adhesion to the stainless steel substrate and no flaking was observed around the scratch tracks. Slight flaking was observed in U1 and U2, which was localized to the area immediately next to the scratch track. This may have been caused by the lack of a Ti adhesion layer in these coatings or due to the stresses applied to the coating during annealing. Given the destructive nature of the scratch test and the high load levels used in this test, all coatings were deemed to show sufficient mechanical resistance for use on food and drinks processing surfaces.

Figure 5. Progressive load scratch tracks of (**a**) T1; (**b**) T2; (**c**) T3; (**d**) U1; (**e**) U2 and (**f**) MC.

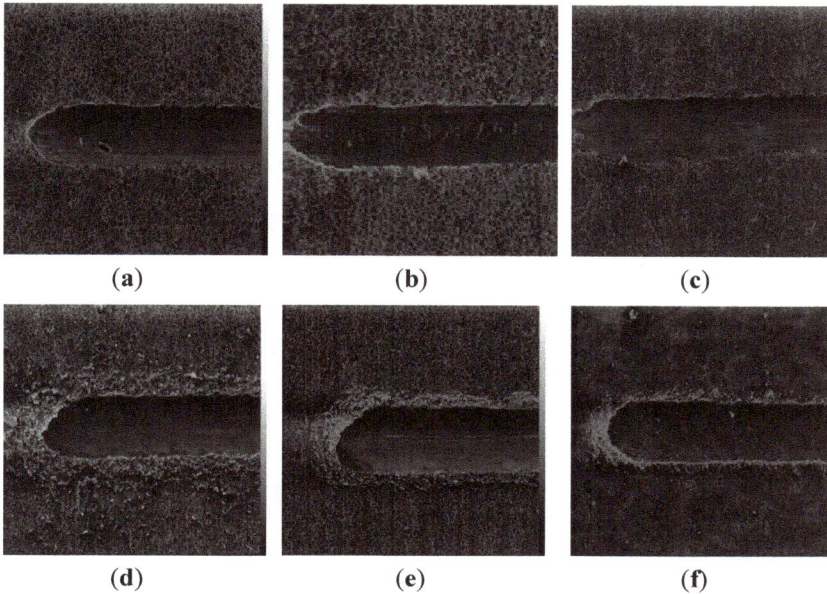

(a)	(b)	(c)

(d)	(e)	(f)

3.2. Properties of the Surfaces after Process Tests

Visual inspection of the coatings after the three months process trial and their comparison with the control surfaces showed that all coatings prepared by magnetron sputtering (T1, T2, T3, U1 and U2) were physically present, although some color changes were apparent (Figure 6). The TiO$_2$ sol coating (MC) appeared to be still present after the process test at Brewery C but was at least partially removed at the other two breweries. It was noticeable that many of the surfaces were heavily soiled, particularly those that had been on trial at Breweries A and B.

Figure 6. Images of coupons after the three month brewery trial. See Table 2 for sample descriptions.

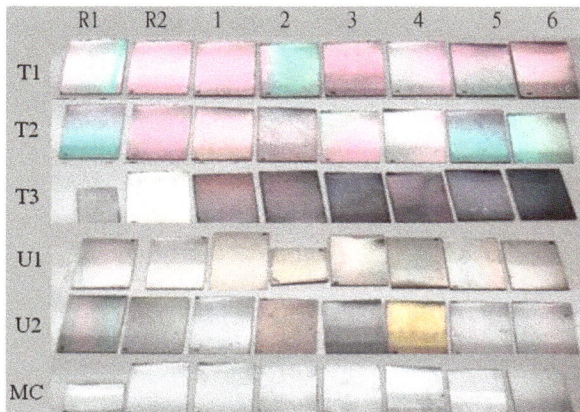

3.3. Mechanical Durability of the Coatings

The results of scratch adhesion tests performed on samples after the process trial confirmed the observations made on the appearance of coatings. Representative results are shown in Figure 7. T1–T3 coatings showed good adhesion with no flaking after the process studies. U1–U2 coatings showed some flaking, which in most cases was confined to the area immediately next to the scratch track. Some of the samples, however, showed a more widespread flaking. This was most likely caused by the lack of a Ti base layer, which can enhance the adhesion of TiO_2 to the stainless steel substrate or alternatively could be a result of the heat treatment. The MC coating from Breweries A and C showed some flaking near the scratch track. Samples removed from Brewery B showed no flaking. EDX analysis of these samples showed a Ti peak which had been greatly reduced compared to that of the control samples, suggesting that the coating had been heavily worn. This could be due to the different cleaning regimes, e.g., chemicals and scrubbing methods used in the different breweries, with some conditions exceeding the chemical and mechanical resistance of the coating.

Figure 7. SEM micrographs showing the scratch tracks of coatings before and after process tests at Brewery C.

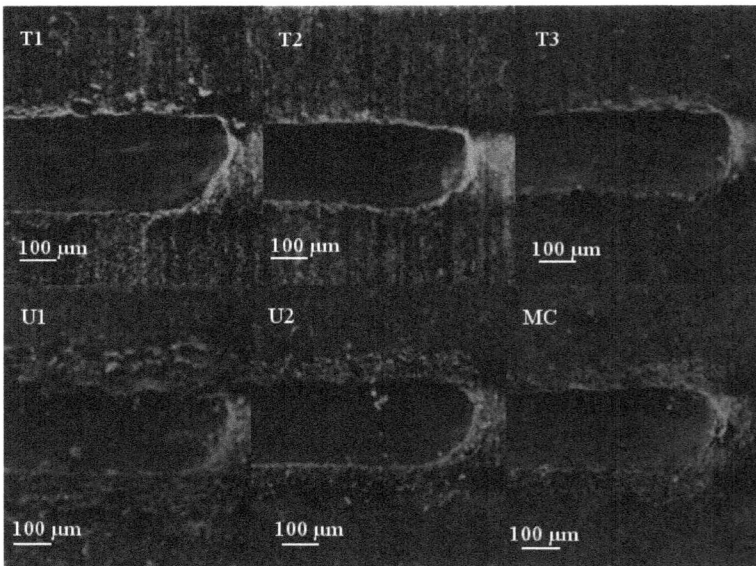

3.4. Composition of the Coatings

EDX results showed that the Ag content in T2 remained fairly constant. T3 showed a high level of Ag leaching possibly caused due to the poor dispersion and segregation of Ag within the coating as was seen from the SEM image of this coating (Figure 3). U2 showed a fairly constant concentration of Mo, except that in the areas where coating had been partially removed, it was not possible to measure the relative concentration of Mo in the coatings due to the weak signal and overlapping of the emission lines from the coating with those from the substrate (Table 3).

Table 3. Concentration of dopant as analysed using EDX (error in the measurements was ±10%).

Coating	As Deposited	Control		Brewery A		Brewery B		Brewery C	
		R1	R2	1	2	3	4	5	6
TiO$_2$-Ag (low Ag) (T2)	0.5	0.5		0.2	0.1	0.3	0.6	0.7	0.1
TiO$_2$-Ag (high Ag) (T3)	30.0	32.0	34.0	1.8	1.6	9.2	1.1	3.6	10.3
TiO$_2$-Mo (U2)	7.0	7.3	7.2	–	–	–	8	8.1	8.0

3.5. Photocatalytic Properties

Figure 8 shows the photocatalytic activity of the coatings under fluorescent and UV irradiation.

Figure 8. Photoactivity of TiO$_2$ and doped TiO$_2$ coatings under UV (blue bars) and fluorescent light (red bars) irradiation. (**a**) T1; (**b**) T2; (**c**) T3; (**d**) U1; (**e**) U2; (**f**) MC.

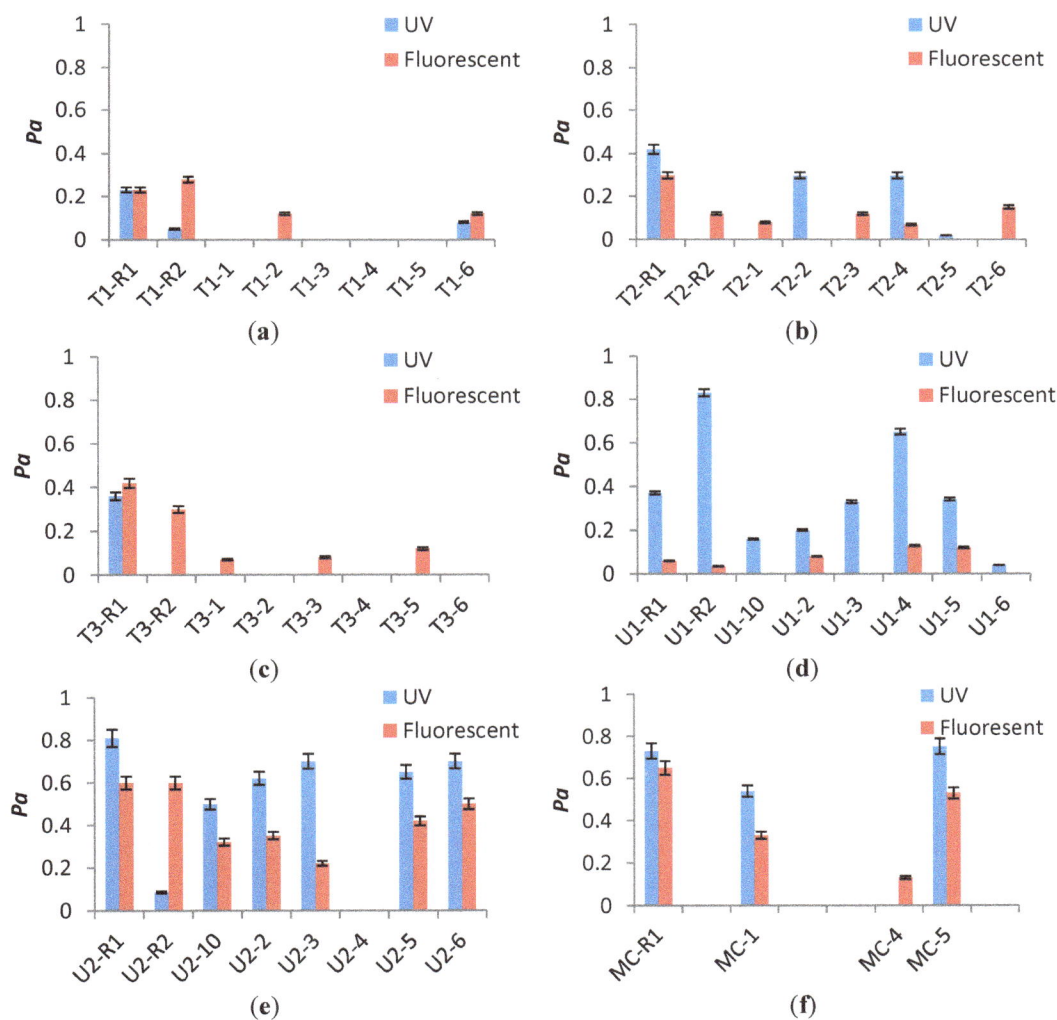

A loss of activity for T1–T3 coatings under UV light following the brewery trials was seen to varying degrees. The lower content TiO_2-Ag surface (T2) retained the most activity with the exception of samples received from Brewery C. A greater loss of photocatalytic properties of the higher doped Ag coatings was seen, possibly due to the leaching of silver during the process studies. The controls also lost activity following three months storage in the dark compared to the as-deposited samples (UV light). Similar results were seen when photocatalytic activity was assessed under fluorescent light. Comparison of the photocatalytic properties of U1 and U2, showed that the addition of Mo to the heat-treated TiO_2 surface increased its photocatalytic activity under UV and fluorescent light and this remained the case following the process studies. Photoactivity was largely retained for Mo-doped surfaces from all breweries with the exception of one of the two samples received from Brewery B. TiO_2 alone retained some of its photoactivity to varying degrees when irradiated with UV, although values between the duplicate samples differ. Less activity was shown under fluorescent light exposure, as expected and controls also showed lower photocatalytic activity compared to the as-deposited samples. Compared to the controls stored in the dark, the MC TiO_2 surfaces retained much of their photocatalytic activity, with the exception of samples received from Brewery B (under UV), where scratch test and EDX results had shown very little coating had been left on the substrate surface after the trial. As a small area of the substrate remained uncoated during the spray coating process, duplicate samples were not available in the case of MC surfaces.

The differences in photocatalytic activities of the surfaces received from the breweries could be due to the position of the samples and the cleaning regimes used. Work by others has shown that canning machines were markedly less prone to accumulation of microorganisms than bottling machines which use recycled glass bottles [1]. Further, it has been suggested that horizontal surfaces were prone to microbial accumulation and should be avoided in constructions as much as possible. Biofilm formation has also been shown to occur on certain surfaces despite daily cleaning and disinfection [1]. Thus, deposits formed by reaction processes or microbes usually cannot be wholly removed with water from stainless steel [29]. Various cleaners may have different success. In a surface test without soil a hypochlorite-based disinfectant was shown to be effective after an exposure of 10 min against all the microbes tested whereas an isopropanol-based cleaning agent was effective against all the vegetative cells tested [30]. In the presence of soil, hypochlorite was effective against *Listeria monocytogenes* and *Pseudomonas aeruginosa* [30]. The nature of clean may also affect efficacy. At 30 and 50 °C water rinsing at the flow velocities investigated could remove up to 85% of a yeast deposit. At a water rinsing temperature of 70 °C, less yeast deposit could be removed overall [3]. If surfaces were soiled with chemical residue and not cleaned sufficiently, it is possible that this may have an effect on photocatalytic activity. Conversely over aggressive cleaners might damage the surface, as noted previously.

3.6. Wettability

Photo-induced hydrophilicity is often associated with photocatalytic TiO_2 coatings [31]. Large differences in the wettability of TiO_2 coatings after irradiation by light or after storage in the dark are believed to be due to the generation of hydrophilic radicals on the TiO_2 surface by the action of light. Measurement of contact angle had been found to be an effective and easy method of detecting the presence of the TiO₂ sol coating, MC. In addition, it is expected that the contact of contaminants with the surface is enhanced in the case of hydrophilic surfaces, resulting in an increase in the effect of the photocatalyst. Thus water contact angle measurements were made on each test coupon listed in Table 3 to help determine the presence and activity of each coating.

Contact angles were firstly measured for the coupons immediately on unpacking (dark), and then after 20 hours irradiation. It was noticeable that many of the coupons were heavily soiled so a portion of each sample was cleaned by wiping with 2-propanol on a soft cloth and then with water. Contact angles were re-measured after 20 h. under UVA light, and again after 6–7 days in the dark. The results are shown in Figures 9–11.

Figures 9–11 show that for most coatings, the effect of light on the wettability was more pronounced in the case of the reference surfaces than those having undergone the processing conditions. This may indicate changes in the coating activity resulting from the exposure to the cleaning chemicals *etc.* used during the processing.

Figure 9. Water contact angle measurements for (**a**) TiO₂ (T1); and (**b**) TiO₂-Ag (low) (T2) coupons after three-month Brewery Trial.

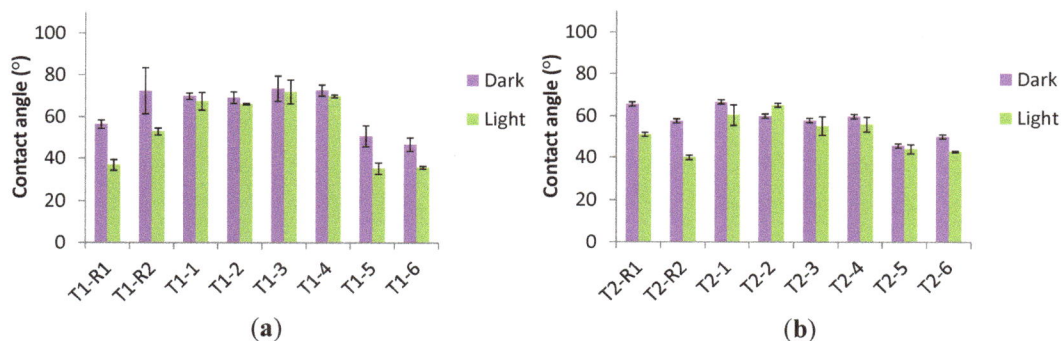

Figure 10. Water contact angle measurements for (**a**) TiO₂-Ag (high) (T3) and (**b**) TiO₂ (U1) coupons after three month Brewery Trial.

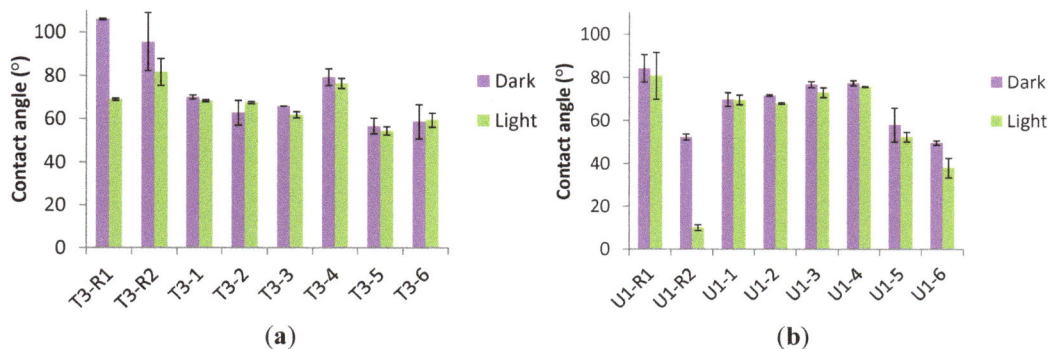

For the TiO₂ sol coating, MC, the two coupons sited at the Brewery C showed similar wettability to the control sample (MC-R2), after cleaning, both in the dark and the light. Visual inspection of the coupons sited at the other two breweries showed that the coating was wholly or partly removed from these coupons, and the contact angle measurements reflect this loss of coating (Figure 11b). Contact angle values on blank stainless steel surfaces after cleaning were 70°–80°.

Figure 11. Water contact angle measurements for (**a**) TiO$_2$-Mo (U2); and (**b**) TiO$_2$ (MC) coupons after three month Brewery Trial (Coated area of MC-R1 was too small to test).

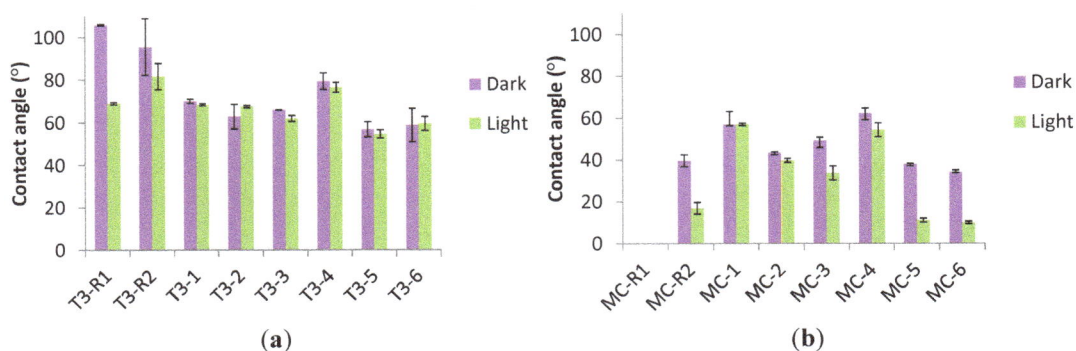

(**a**) (**b**)

4. Conclusions

TiO$_2$ coatings were deposited either using reactive magnetron sputtering, both with and without subsequent heat treatment, or prepared by spray coating. Photocatalytic activity, determined by methylene blue degradation, was high under UV irradiation. The coatings were also active under fluorescent irradiation. Doping of magnetron sputtered TiO$_2$ with Ag- (0.5 at%) and Mo- (7 at%) increased the activity under fluorescent light. High Ag loading (~30%) had a detrimental effect on the fluorescent light induced photoactivity, possibly due to the replacement of Ti atoms in the TiO$_2$ matrix with Ag. The coatings were placed in three different breweries for three months. The magnetron sputtered TiO$_2$ surfaces which had not undergone heat treatment showed the best mechanical resistance, whilst the spray coated TiO$_2$ and Mo-TiO$_2$ showed the best retention of photoactivity. Irradiation of the coatings resulted in an increase in wettability, but the spray-coated TiO$_2$ was the only coating showing light-induced hydrophilicity after the process trial.

This work presented the potential of magnetron sputtered TiO$_2$ and doped TiO$_2$ coatings for surfaces used in food and beverage processing where there is a requirement for robust coatings. Selection of the optimum deposition parameters and dopants can lead to coatings which retain photoactivity and are durable in harsh processing conditions. The use of spray coatings is preferred on surfaces which do not experience severe mechanical wear and abrasion.

Acknowledgments

This work was part of the MATERA+ Project "Disconnecting", refer MFM-1855, Project No. 620015. Funding from the Technology Strategy Board, the UK's innovation agency and Finnish Funding Agency for Technology and Innovation (Tekes) is gratefully acknowledged. The authors would also like to thank Oy Panimolaboratorio - Bryggerilaboratorium Ab (PBL Brewing Laboratory) for their collaboration in this work.

Author Contributions

Parnia Navabpour, Kevin Cooke, Soheyla Ostovarpour, Peter Kelly, Joanna Verran and Kathryn Whitehead provided the planning and experimental work on magnetron sputtered coatings and the sections of the paper on these coatings. Teer Coatings authors also carried out the scratch tests and SEM work on all coatings and were responsible for writing the sections of the manuscript on these results. Manchester Metropolitan University authors were responsible for the photocatalytic tests, as well as the manuscript sections on these results. Carin Tattershall and Claire Hill carried out the work on MC coating and contact angle analysis and wrote the article sections on these results. Outi Priha and Mari Raulio were responsible for coordinating and performing the process tests. All authors contributed in the discussion and improvement of the paper. Parnia Navabpour coordinated the writing of the overall manuscript.

Conflicts of Interest

The authors declare no conflict of interest.

References

1. Storgårds, E.; Tapani, K.; Hartwall, P.; Saleva, R.; Suihko, M. Microbial attachment and biofilm formation in brewery bottling plants. *J. Am. Soc. Brew. Chem.* **2006**, *64*, 8–15.
2. Priha, O.; Laakso, J.; Levänen, E.; Kolari, M.; Mäntylä, T.; Storgårds, E. Effect of photocatalytic and hydrophobic coatings on brewery surface microorganisms. *J. Food Prot.* **2011**, *11*, 1788–1989.
3. Goode, K.R.; Asteriadou, K.; Fryer, P.J.; Picksley, M.; Robbins, P.T. Characterising the cleanign mechanisms of yeast and the implicaitons for Cleaning In Place (CIP). *Food Bioprod. Proc.* **2010**, *88*, 365–374.
4. Timke, M.; Wang-Lieu, N.Q.; Altendorf, K.; Lipski, A. Identity, beer spoiling and biofilm forming potential of yeasts from beer bottling plant associated biofilms. *Antonie Van Leeuwenhoek* **2008**, *93*, 151–161.
5. Fornalik, M. Biofouling and process cleaning: A practical approach to understanding what is happening on the walls of your pipes. *Master Brew. Assoc. Am. Tech. Q.* **2008**, *45*, 340–344.
6. Storgårds, E. Process Hygiene Control in Beer Production and Dispensing. Available online: http://www.vtt.fi/inf/pdf/publications/2000/P410.pdf (accessed on 8 July 2014).
7. Rezić, T.; Rezić, I.; Blaženović, I.; Šantek, B. Optimization of corrosion processes of stainless steel during cleaning in steel brewery tanks. *Mater. Corros.* **2013**, *64*, 321–327.
8. Farahani, N.; Kelly, P.J.; West, G.; Ratova, M.; Hill, C.; Vishnyakov, V. Photocatalytic activity of reactively sputtered and directly sputtered titania coatings. *Thin Solid Films* **2011**, *520*, 1464–1469.
9. Caballero, L.; Whitehead, K.A.; Allen, N.S.; Verran, J. Inactivation of E.coli on immobilized TiO_2 using fluorescent light. *J. Photochem. Photobiol. A* **2009**, *202*, 92–98.

10. Bandaranayake, K.M.P.; Indika Senevirathna, M.K.; Prasad Weligamuwa, P.M.G.M.; Tennakone, K. Dye-sensitized solar cells made from nanocrystalline TiO$_2$ films coated with outer layers of different oxide materials. *Coord. Chem. Rev.* **2004**, *248*, 1277–1281.

11. Sung, Y.M.; Kim, H.J. Sputter deposition and surface treatment of TiO$_2$ films for dye-sensitized solar cells using reactive RF plasma. *Thin Solid Films* **2007**, *515*, 4996–4999.

12. Pakdel, E.; Daoud, W.A.; Wang, X. Self-cleaning and superhydrophilic wool by TiO$_2$/SiO$_2$ nanocomposite. *Appl. Surf. Sci.* **2013**, *275*, 397–402.

13. Samal, S.S.; Jeyaraman, P.; Vishwakarma, V. Sonochemical Coating of Ag-TiO$_2$ Nanoparticles on Textile Fabrics for Stain Repellency and Self-Cleaning- The Indian Scenario: A Review. *J. Miner. Mater. Charact. Eng.* **2010**, *9*, 519–525.

14. Hájková, P.; Špatenka, P.; Krumeich, J.; Exnar, P.; Kolouch, A.; Matoušek J.; Koči, P. Antibacterial effect of silver modified TiO$_2$/PECVD films. *J. Eur. Phy. D* **2009**, *54*, 189–193.

15. Miao, L.; Tanemura, S.; Kondo, Y.; Iwata, M.; Toh, S.; Kaneko, K. Microstructure and bactericidal ability of photocatalytic TiO$_2$ thin films prepared by rf helicon magnetron sputtering. *Appl. Surf. Sci.* **2004**, *238*, 125–131.

16. Tanemura, S.; Miao, L.; Wunderlich, W.; Tanemura, M.; Mori, Y.; Toh, S.; Kaneko, K. Fabrication and characterization of anatase/rutile-TiO$_2$ thin films by magnetron sputtering: A review. *Sci. Technol. Adv. Mater.* **2005**, *6*, 11–17.

17. Jiang, D.; Zhang, S.; Zhao, H. Photocatalytic Degradation Characteristics of Different Organic Compounds at TiO$_2$ Nanoporous Film Electrodes with Mixed Anatase/Rutile Phases. *Environ. Sci. Technol.* **2007**, *41*, 303–308.

18. Andersson, M.; Österlund, L.; Ljungström, S.; Palmqvist, A. Preparation of nanosize anatase and rutile TiO$_2$ by hydrothermal treatment of microemulsions and their activity for photocatalytic wet oxidation of phenol. *J. Phy. Chem. B* **2002**, *106*, 10674–10679.

19. Yin, H.; Wada, Y.; Kitamura, T.; Kambe, S.; Murasawa, S.; Mori, H.; Sakata, T.; Yanagida, S. Hydrothermal synthesis of nanosized anatase and rutile TiO$_2$ using amorphous phase TiO$_2$. *J. Mater. Chem.* **2001**, *11*, 1694–1703.

20. Wasa, K.; Kitabatake, M.; Adachi, H. *Thin Film Materials Technology: Sputtering of Control Compound Materials*; William Andrew Inc.: Norwich, CT, USA, 2004.

21. Laing, K.; Hampshire, J.; Teer, D.G.; Chester, G. The effect of ion current density on the adhesion and structure of coatings deposited by magnetron sputter ion plating. *Surf. Coat. Technol.* **1999**, *112*, 177–180.

22. Ohtani, T.; Ogawa, Y.; Nishimoto, S. Photocatalytic Activity of Amorphous−Anatase Mixture of Titanium(IV) Oxide Particles Suspended in Aqueous Solutions. *J. Phy.Chem. B* **1997**, *101*, 3746–3752.

23. Yu, J.; Zhao, X.; Du, J.; Chen, W. Preparation, microstructure and photocatalytic activity of the porous TiO$_2$ anatase coating by sol-gel processing. *J. Sol Gel Sci. Technol.* **2000**, *17*, 163–171.

24. Nam, H.; Amemyima, T.; Murabayashi, M.; Itoh, K. Photocatalytic Activity of Sol-Gel TiO$_2$ Thin Films on Various Kinds of Glass Substrates: The Effects of Na$^+$ and Primary Particle Size. *J. Phy. Chem. B* **2004**, *108*, 8254–8259.

25. Kubacka, A.; Colón G.; Fernández-García, M. Cationic (V, Mo, Nb, W) doping of TiO$_2$–anatase: A real alternative for visible light-driven photocatalysts. *Catal. Today* **2009**, *143*, 286–292.

26. Onifade, A.A.; Kelly, P.J. The influence of deposition parameters on the structure and properties of magnetron-sputtered titania coatings. *Thin Solid Films* **2006**, *494*, 8–12.

27. Stallard, J.; Poulat, S.; Teer, D.G. The study of the adhesion of a TiN coating on steel and titanium alloy substrates using a multi-mode scratch tester. *Tribol. Int.* **2006**, *39*, 159–166.

28. Ratova, M.; Kelly, P.J.; West, J.T.; Iordanova, I. Enhanced properties of magnetron sputtered photocatalytic coatings via transition metal doping. *Surf. Coat. Technol.* **2013**, *228*, S544–S549.

29. Christian, G.K.; Fryer, P.J.; Liu, W. How hygiene happens: Physics and chemistry of cleaning. *Int. J. Dairy Technol.* **2006**, *59*, 76–84.

30. Grönholm, L.; Wirtanen, G.; Ahlgren, K.; Nordström, K.; Sjöberg, A.-M. Screening of antimicrobial activities of disinfectants and cleaning agents against foodborne spoilage microbes. *Z. Lebensm. Unters. Forshung A* **1999**, *208*, 289–298.

31. Fujishima, A.; Rao, T.N.; Tryk, D.A. Titanium dioxide photocatalysis. *J. Photochem. Photobiol. C* **2000**, *1*, 1–21.

Photocatalytic Properties of Nb/MCM-41 Molecular Sieves: Effect of the Synthesis Conditions

Caterine Daza Gomez and Jorge Enrique Rodriguez-Paez

Abstract: The effect of synthesis conditions and niobium incorporation levels on the photocatalytic properties of Nb/MCM-41 molecular sieves was assessed. Niobium pentoxide supported on MCM-41 mesoporous silica was obtained using two methods: sol-gel and incipient impregnation, in each case also varying the percentage of niobium incorporation. The synthesized Nb-MCM-41 ceramic powders were characterized using the spectroscopic techniques of infrared spectroscopy (IR), Raman spectroscopy, X-ray diffraction (XRD), and transmission electron microscopy (TEM). The photodegradation capacity of the powders was studied using the organic molecule, methylene blue. The effect of both the method of synthesis and the percentage of niobium present in the sample on the photodegradation action of the solids was determined. The mesoporous Nb-MCM-41 that produced the greatest photodegradation response was obtained using the sol-gel method and 20% niobium incorporation.

Reprinted from *Coatings*. Cite as: Gomez, C.D.; Rodriguez-Paez, J.E. Photocatalytic Properties of Nb/MCM-41 Molecular Sieves: Effect of the Synthesis Conditions. *Coatings* **2015**, *5*, 511-526.

1. Introduction

A photocatalyst is a material able to absorb light efficiently and thereby induce a chemical reaction [1]. In the 1970s, the photo-response of solid materials became a very important research topic due to the potential applications of this phenomenon in new technologies. This led to heterogeneous photocatalysis encouraging unique developments with applications in alternative energy, organic synthesis and environmental treatment [2], with semiconductor compounds being most widely used for this purpose. Of these, TiO_2 and TiO_2-based materials have by far been the most attractive economically: they are authoritative photocatalysts also because they are nontoxic and offer among other advantages ready availability, chemostability, and reusability. Other inorganic semiconductors have been successfully used as photocatalysts, principally in environmental heterogeneous photocatalysis. These include CdS, WO_3, SnO_2, α-Fe_2O_3, $AgNbO_3$ and $SrTiO_3$ [3]. Basically, when such semiconductor photocatalysts absorb light, electronic excitation occurs that can be used in chemical or electrical processes.

During all this time, knowledge in disciplines such as photochemistry, catalysis and semiconductor physics has grown and allowed a deeper understanding of the phenomenon of photocatalysis, which depends mainly on the interaction of radiation with the solid and its surface reactivity [4]. The basic concepts underlying the phenomenon of photocatalysis have been obtained from studies done on photocatalytic systems employing microcrystalline (bulk) semiconductor photocatalysts. Recently, nanotechnology has opened up exciting new possibilities to develop more efficient photocatalysts [5]. This modernization of photocatalysis and nanophotocatalysis has been encouraged by progress that has been made in the physics and chemistry of nanoscale semiconductor particles exhibiting quantum size effects, *i.e.*, a dependence on the fundamental properties of these nanosized particles. The phenomenon of quantum confinement, which these nanoparticles exhibit, favors changes in the electronic, optical, photochemical and photocatalytic properties of nanocrystalline semiconductors [6,7].

Much work has been published in the last two decades related not only to colloidal semiconductors but to other types of nanostructure materials, including: nanocrystalline powders, coupled nanocomposites, nanocrystalline films, mesoporous semiconductors, *etc*. Clearly, although nanoparticle semiconductors have high surface energy compared to bulk semiconductors, they tend to reduce surface tension by aggregation. In order to prevent or delay aggregation and counter the coarsening of the nanocrystals, a stabilizer or surfactant is added during synthesis that is adsorbed on them, thereby reducing surface tension and creating a steric or electrostatic barrier. In addition, this characteristic shown by the nanoparticles can be used to create a new class of photocatalysts with a porous structure [8]. For this, surfactant type compounds are introduced during synthesis in suitable concentrations, not so much to prevent aggregates from forming but to form a nanoporous structure in the aggregates. As photocatalysts, mesoporous materials have a number of advantages, including: (1) a high value of specific surface area and nanometer pore size; (2) prolonged substrate retention; (3) migration of photogenerated charge carriers through the adjacent semiconductor nanoparticles, and an accumulation at their points of contact, and (4) high possibility of repeat light reflectance encouraging a greater absorption of light [9]. The benefits of these mesoporous structures are illustrated by the titania nanophotocatalysts in the photocatalytic oxidation of dyes [10], oxidative disruption of the cell membranes of *E. coli* bacteria [11] and carbon dioxide reduction to methanol [12], among others. Specifically, mesoporous ZnO [13], CeO_2 [14] and Co_3O_4 [15] have all exhibited high photocatalytic activity in the destruction of dyes when compared with non-porous nanopowders.

Specifically, considering mesoporous niobium compounds, in one of the first works where the mesoporous niobium-containing silicate of MCM-41 was synthesized, the results of X-ray diffraction and transmission electron microscopy confirmed that this material exhibited an ordered mesoporous structure that could be altered by applying external pressure (~50 MPa) [16]. According to these researchers, the niobium-containing MCM-41 structure is mechanically less stable than pure silica MCM-41. The mesoporous molecular sieves were then modified with ammonium and copper cations and tested in several catalytic reactions [17]. An important result of this study showed that mesoporous NbMCM-41 was an attractive catalyst for the selective oxidation of thioethers to sulphoxides with H_2O_2 owing to its high activity and to the presence of Nb-O$^-$ species in the framework [18]. Recently, mesoporous Nb_2O_5 was used as a photocatalyst instead of the nanocrystalline niobium oxide compound, and its quantum yield of hydrogen evolution from aqueous methanol solutions increased by a factor of 20 [19]. When the mesoporous Nb_2O_5 samples were loaded with nanoparticles of Pt, Au, Cu and NiO, their photocatalytic activity increased such that the rate of H_2 evolution from an aqueous methanol solution under UV irradiation changed, being higher for the Pt-loaded niobium oxide catalysts [20]. Another important application of Nb_2O_5 photocatalysts has been the oxidation of organic compounds in aqueous media [21]. This involves natural and synthetic Nb_2O_5 being placed initially in the presence of hydrogen peroxide and methylene blue [22] and in another study niobium pentoxide was modified by doping it with

120

molybdenum or tungsten and then treated with H_2O_2, significantly increasing the photocatalytic activity of this compound in the oxidation of methylene blue dye [23].

This article reports obtaining Nb/MCM-41 molecular sieves by using two synthesis methods: the traditional incipient impregnation method and the sol-gel process. After characterizing the synthesized powders, their photodegradation capacity was evaluated using methylene blue as a pattern compound. Finally, also with regard to their photodegradation capacity, the effect of the synthesis method for the Nb/MCM-41 photocatalysts was determined, together with that of the niobium content in the different samples.

2. Experimental Procedure

2.1. Synthesis of Different Molecular Sieves

2.1.1. Nb_2O_5 Synthesis Supported on MCM-41 through the Sol-Gel Method (Nb-MCM-41-Solgel)

A total of 8.8 g of N-cetyl-N,N,N-trimethylammonium bromide (Merck, Darmstadt, Germany) were dissolved in a solution of 208 mL of deionized water and 96 mL of aqueous ammonia solution (25% w/w, Merck, Darmstadt, Germany) at 35 °C. A total of 40 mL of TEOS (98% purity, Aldrich®, San Luis, MO, USA) and an aqueous solution of ammonium tris(oxalate) complex of niobium(V) (CBMM, Araxá, Minas Gerais, Brazil) were slowly added under agitation to this clear solution, in 0.1 M of oxalic acid; the amounts of niobium oxalate aggregated were 0.4756, 0.9512, and 1.4268 g to obtain niobium percentages of 10%, 20%, and 30%, respectively, in the mesoporous silica material. Thereafter, the system was agitated for three hours and the gel aged at room temperature for 24 h in a sealed container. The precipitate was filtered, washed with 600 mL of deionized water, and air dried at room temperature. Finally, to eliminate the organic matter, it was calcined at 550 °C for eight hours. The procedure previously described was carried out taking into account its simplicity, reproducibility, and the infrastructure currently available in the laboratory. Certainly, the aim in future would be to incorporate further stages, used to good effect by other researchers, in order to optimize the characteristics of the end product.

2.1.2. Nb_2O_5 Synthesis Supported on MCM-41 through the Incipient Humidity Impregnation Method (Nb/MC-41-ImpHum)

Impregnation was carried out using 10 mL of an aqueous solution of niobium oxalate in 0.1 M of oxalic acid with which 1.1034 g of the MCM-41 material were impregnated, drop by drop, until the whole solid was humid; the amount of niobium oxolate aggregated was 0.4756, 0.9512, and 1.4268 g to obtain niobium percentages of 10%, 20%, and 30%, respectively, in the mesoporous silica material. The humid solid was air dried at 60 °C and, thereafter, calcined at 550 °C for 6 h.

2.2. Characterization of Photocatalysts

The synthesized solids were micro-structurally characterized using: X-ray diffraction (XRD), Fourier Transformed Infrared Spectroscopy (FTIR), Raman spectroscopy, and transmission electron microscopy (TEM). To perform the IR spectroscopy analysis, the solid was homogenized with spectroscopic grade potassium bromide (99%, Fischer, Pittsburgh, USA) in an agate mortar. The mixture was subjected to pressure through a 318 stainless steel die until forming a translucent pellet. The sample was analyzed with a Nicolet IR-200 infrared spectrophotometer equipped with EZOMINIC 32 software. A total of 32 scans were conducted at a resolution of 16 cm^{-1}/s. For the XRD study of the powder samples, a PANalytical X'Pert Pro X-ray diffractometer was used with Bragg Brentano geometry, equipped with $CuK\alpha$ (λ = 1.5406 Å) radiation source, operating with 45 mA current and voltage of 45 kV. Powder X-ray diffraction patterns (PXRD) were registered within the 1 to 10° interval, as well as between 10° and 70°, in 2θ, at 0.5 °/min scan rate. The samples were observed with transmission electron microscopy (Jeol 1200 EX, JEOL, Pleasanton, CA, USA) with an electron beam electric potential acceleration of 80 kV. Raman spectra were obtained through EZRaman-N and ProRaman-L-905 (Enwave Optronics, Irvine, USA) Raman analyzers coupled to a Leica DM300 microscope (with a Leica objective that has a magnification/numerical aperture ratio of 40×/0.65), using excitation laser sources of 905 nm (maximum power ~400 mW). The adsorption/desorption of nitrogen was carried out at −196 °C in a Micromeritics ASAP 2010 (micromeritics, USA). The samples were previously treated under high vacuum at 150 °C for 12 h. Finally, elemental composition was determined by SEM-EDX in a Stereoscan 440 Leica microscope (Leica Microsystems, Mannheim, Germany) equipped with an energy dispersive X-ray (EDX) elemental analysis system.

2.3. Photocatalytic Activities

2.3.1. Adsorption Kinetics

To determine the characteristics of the dye adsorption equilibrium on the molecular sieves, a study was conducted on their adsorption kinetics. For this purpose, 10 mg were taken of each of the solids synthesized and these were dispersed in 200 mL of methylene blue solution with a concentration of 50 ppm, at room temperature, under continuous agitation and complete darkness. Thereafter, every 10 min, aliquots of this suspension were extracted and their spectra were taken, considering the maximum absorbance of the methylene blue at 665 nm using a UV-Vis spectrophotometer (Spectronic Genesys 6); this procedure was repeated over a total time of 60 min. Then, through mathematical calculations and using the calibration curve method, the quantity in milligrams of methylene blue milligrams adsorbed on the solids was determined; the curve for the "q" coefficient (mg of methylene blue adsorbed/mg of the photocatalyst) as a function of time was obtained. This permitted determining the time interval the system should be kept in continuous agitation prior to turning on the UV lamps to determine the photodegradation capacity of suspended Nb/MCM-41 solids.

2.3.2. Photodegradation Effect

To establish the photodegradation capacity of the synthesized solids, methylene blue powder (C16H18N3Cl S) was taken as the reference organic molecule to be degraded. This powder was dissolved in distilled water at a concentration of 55 ppm and 200 mL were taken from the resulting solution and poured into a 250-mL precipitation glass. Then, 10 mg of the synthesized powders were weighed and added to the methylene blue solution. The mixture was agitated for the time determined from the adsorption kinetics to eliminate or reduce as much as possible the adsorption effect, and then transferred to a "solar simulator" constructed in the laboratory. The suspension was placed in the simulator and the UV lamps (Phillips TUV 15 W lamps, maximum intensity at 254 nm, Phillips, Lausanne, Switzerland) turned on to begin the photodegradation process. During this process, the system was maintained at a continuous agitation of 500 rpm/min, taking samples periodically every 5 min. The samples were placed in a small quartz container and placed in the UV-Vis spectrophotometer to obtain the system's absorbance value at a wavelength of 665-nm; identical steps were followed for the photolysis test, but using only the methylene blue solution. Upon determining the absorbance, the suspension was again poured into the precipitation glass. This process was repeated every 5 min, until minute 20, and then every 10 min until minute 60. Finally, the methylene blue percentage degradation over time curve was obtained with respect to the calibration curve. The photodegradation effect was determined by employing a laboratory designed photo-reactor. Monitoring of the disappearance of methylene blue in the solution was carried out using UV-Vis spectroscopy.

3. Results and Discussion

3.1. Characterization of Solids

The FTIR spectrum of Nb-MCM-41 materials is shown in Figure 1, where bands are observed between 1400 and 400 cm^{-1} due to the fundamental vibrations of the mesoporous structure. A large band around 3450 cm^{-1} and a band at ~1630 cm^{-1} correspond to O–H stretching and surface water [24]. The bands at ~1230, ~1080, and ~810 cm^{-1} are assigned to Si–O symmetric and asymmetric stretch vibrations; the band at ~460 cm^{-1} is characteristic of silica compounds and corresponds to O–Si–O group stretching [25]. This infrared spectrum was the same independent of the procurement method and percent incorporation of niobium.

Raman spectroscopy characterization permits to determine, to a certain degree, the incorporation of niobium pentoxide onto the MCM-41 mesoporous structure. The lack of a neat Raman band in the spectrum at ~680 cm^{-1}—corresponding to niobium oxide polyhedra symmetric stretching modes—or bands between 200 and 300 cm^{-1}—assigned to flexion modes of Nb–O–Nb bonds [26]—would indicate a lack of Nb$_2$O$_5$ crystalline nanoparticles on the silica structure. Bearing this in mind, it can be observed in Figure 2 that the Raman band at ~680 cm^{-1} is present in most of the solids analyzed, except for the mesoporous solids with 10% niobium synthesized through the sol-gel method, Figure 2d, and with 20% niobium obtained through the incipient impregnation method, Figure 2b, which indicates that most of the niobium added was incorporated into the silica structure [26]. However, in the other solids the presence of this band indicates that not all the niobium added was incorporated into the silica, but that a certain percentage of it was leached. Though the increase occurring in the background of the Raman spectra, Figure 2b,d, can be associated with the fluorescence of organics present in the sample [27], due to the presence of carbon in the sample as it was found in the images of SEM-EDS (Figure 3), however the carbon content is small.

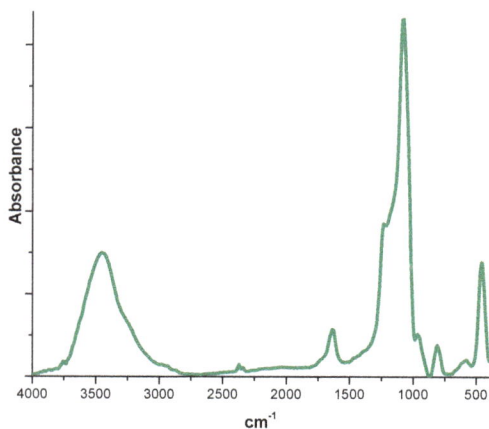

Figure 1. Infrared spectrum of Nb-MCM-41 sample.

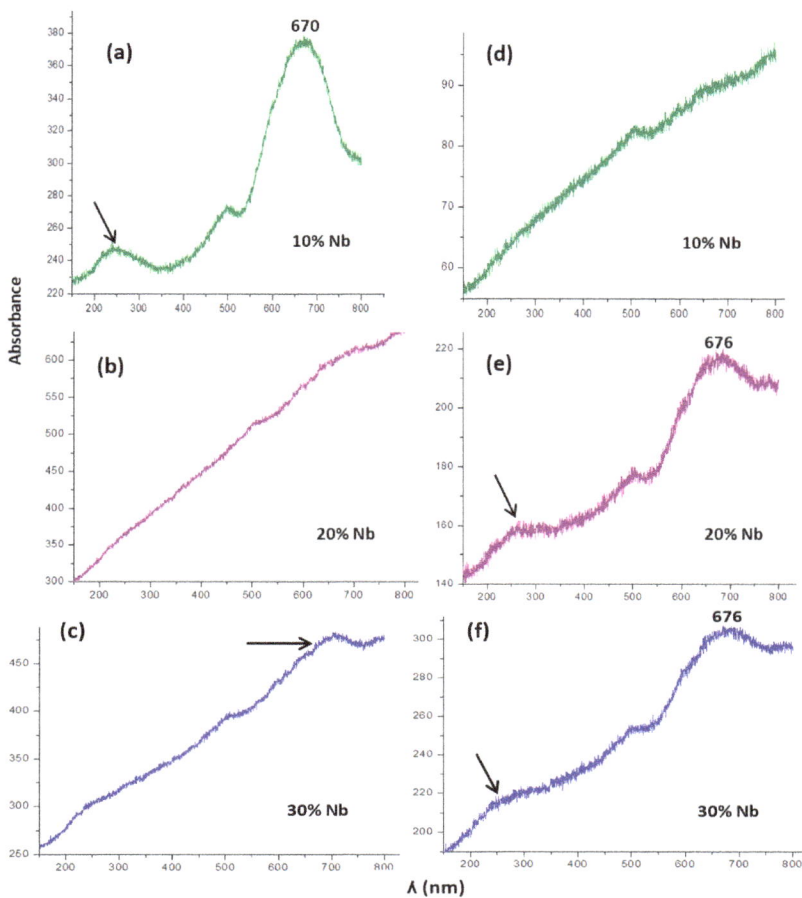

Figure 2. Raman spectrum of Nb-MCM-41 through the incipient impregnation method (**a**), (**b**), and (**c**), and through the sol-gel method (**d**), (**e**), and (**f**).

In order to verify the niobium concentration on the surface of the MCM-41 mesoporous structure, EM-EDS analysis was conducted, shown in Figure 3.

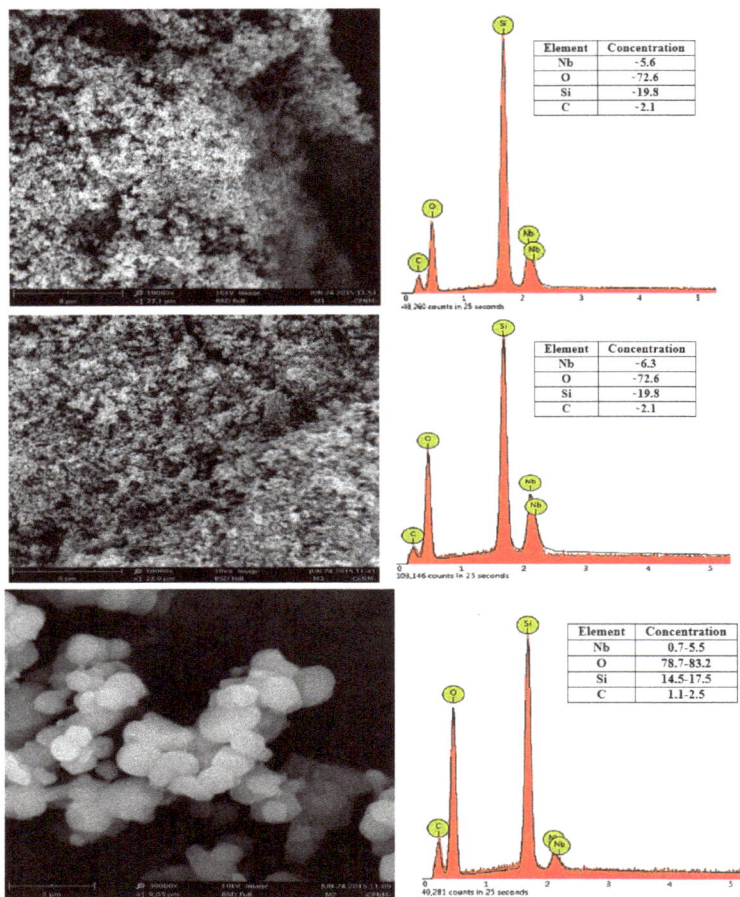

Figure 3. SEM-EDS photograph of Nb-MCM-41 (**a**) 20%-Nb-MCM-41, (**b**) 30%-Nb-MCM-41 powders obtained through the sol-gel method and (**c**) 20%-Nb-MCM-41 powder obtained through the incipient impregnation method.

The Nb-MCM-41 powders obtained through the sol-gel method showed a greater distribution of niobium on the surface. In those obtained by the incipient impregnation method, however, the niobium distribution was not homogeneous.

The Nb-MCM-41 materials obtained using the incipient impregnation method (Figure 4) produces an X-ray diffractogram characteristic of MCM-41, which contains four peaks at low angles, as noted in the figure; the first of these is the most intense peak, appearing around $2\theta = 2°$, corresponding to Miller index (100). Other lower-intensity reflections appear between $3° < 2\theta < 10°$, corresponding to Miller indexes (110) and (200), which verifies a hexagonal symmetry of the structure [24,25,27]; the X-ray diffractogram is the same independent of the percentage of niobium incorporated. This result is similar to that obtained by Ziolek and Nowak [16].

Furthermore, observing the complete diffractogram between 10° and 90°, Figure 4b, similar for all samples impregnated with Nb, two ridges are observed in the regions where, normally, the peaks characteristic of Nb_2O_5 are located, so that these samples may contain crystals of this oxide forming. Regarding the solids obtained using the sol-gel method, these produce a diffractogram (Figure 5) that differs from the typical MCM-41structure, given that it does not show this structure's characteristic peaks. However, these solids show peaks at low angles, which could indicate that a mesoporous structure is being obtained, but not an MCM-41 type structure.

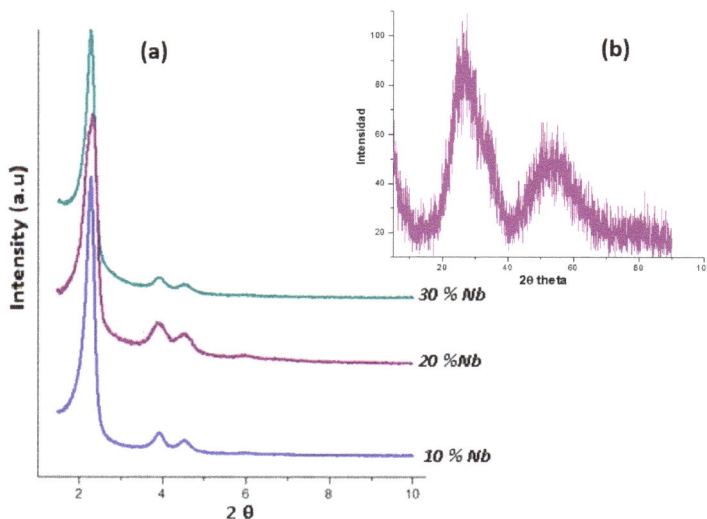

Figure 4. X-ray diffractogram of Nb-MCM-41 obtained using the incipient impregnation method to (**a**) in the region of low angles, and (**b**) in the high-angle region.

These diffraction patterns are different from those obtained by Ziolek and Nowak [16] in that the mesoporous Nb-MCM-41 synthesized by the sol-gel method should show a different pore structure. Given the similarity of the XRD patterns of Figure 5 to the small-angle XRD patterns of the mesoporous Nb samples synthesized by Chen *et al.* [19] that showed obvious peaks between 0.58 and 1.58 and no other peak at higher degrees, it can be concluded that these came from a less ordered mesoporous structure and that the d value of each mesoporous sample would lie between 7.7 and 14.9 nm.

Moreover, observing the full diffractogram between 10° and 90°, Figure 5b, similar for all the samples synthesized using different concentrations of Nb precursor, this is similar to that of Figure 4b, so that it is possible these samples also contain in their structure some forming Nb_2O_5 crystals.

To corroborate the mesoporous structure, the N_2 adsorption and desorption isotherms are shown in Figure 6. This analysis allowed to determine of the Nb-containing mesoporous MCM-41 solid along with large surface areas after the incorporation of niobium to the structure of the silica. As the load increases niobium, surface area decreases as expected (Table 1).

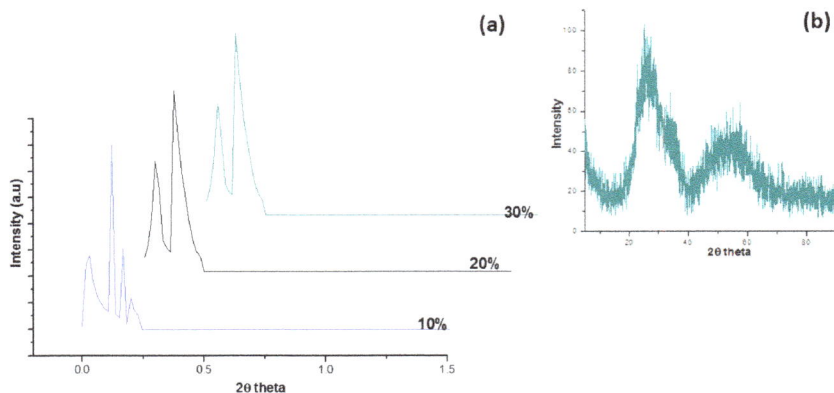

Figure 5. X-ray diffractogram of Nb-MCM-41 obtained through the sol-gel method with 10%, 20% and 30% of niobium.

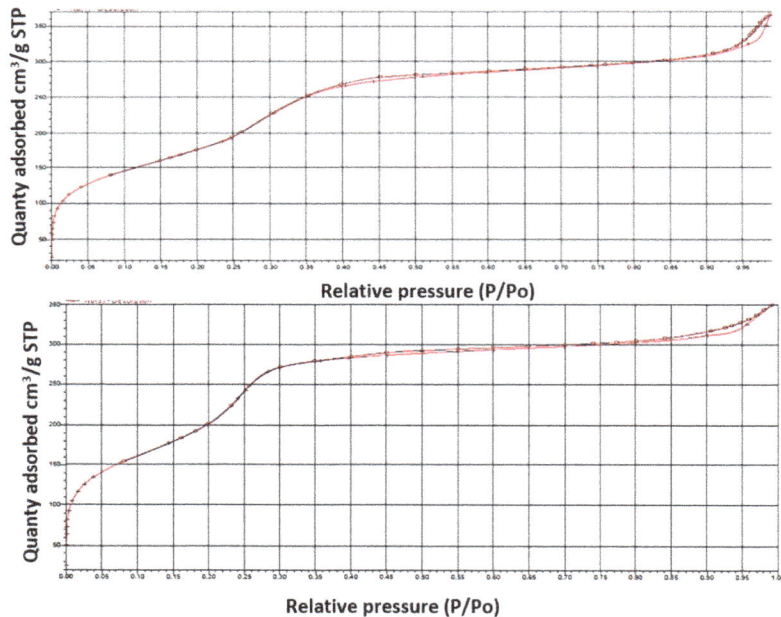

Figure 6. N_2 adsorption and desorption isotherms of 20% Nb-MCM-41 obtained through (**a**) sol-gel and (**b**) incipient impregnation methods, respectively.

However, although the samples exhibited a mesoporous phase, which was corroborated using DRX at low angles, their hysteresis loop is different compared to that of MCM-41, possibly through the formation of species on the surface of the pores from the niobium, which leads to a "plugging" of the channels, and which was reflected in the narrowness of the hysteresis loop area [28], as well as decreasing both the surface area and pore size. This plugging would be generated by the synthesis methods used to obtain the samples of interest. In the case of impregnation of the MCM-41, it is very likely that niobium species are deposited, during the impregnation process, in the pores of the substrate (MCM-41), plugging them and causing a reduction in the hysteresis loop (see Figure 6b). In the case of the samples synthesized by sol-gel, after forming the micelles and adding the organic Nb precursor (ammonium oxalate), it is possible that due to its nature, some of it is distributed within the micelles (lyophobic area) and remains there throughout the process of synthesis, leading finally to plugging of the pores and thus a reduction of the hysteresis loop (Figure 6a). Clearly, on increasing the concentration of Nb precursor, a greater plugging of the pores can be expected, and a reduction in their size, as shown in Table 1.

Table 1. Textural properties of mesoporous solids synthesized in this work.

Sample	Mass (g)	BET Surface Area (m²/g)	Correlation Coefficient	Pore Volume (cm³/g)	Pore Size (nm)	Pore Type
20% Nb-MCM-41 sol-gel method	0.3896	684.1391 ± 24.5850	0.9967814	0.618766	2.61	Mesopore
30% Nb-MCM-41 sol-gel method	0.0659	496.8287 ± 10.1249	0.9993733	0.458849	2.42	Mesopore
20% Nb-MCM-41 incipient impregnation method	0.0441	836.4673 ± 41.6990	0.9924810	0.610336	2.44	Mesopore

The TEM micrographs for the Nb-MCM-41 solids obtained through the incipient impregnation method (Figure 7) should not present optimal pore distribution, mainly because these solids were calcined again following impregnation with the niobium precursor, which would generate some loss of organized structure.

Figure 7. TEM micrographs corresponding to Nb-MCM-41 mesoporous solids obtained using the incipient impregnation method with 30% niobium.

The TEM micrographs corresponding to Nb-MCM-41 mesoporous solids synthesized using the sol-gel method are also shown in Figure 8, illustrating that pore distribution is uniform, which indicates that the pore size for these solids ranges between 2 and 10 nm. These transmission electron micrographs are typical of mesoporous structures, according to several reports [29].

128

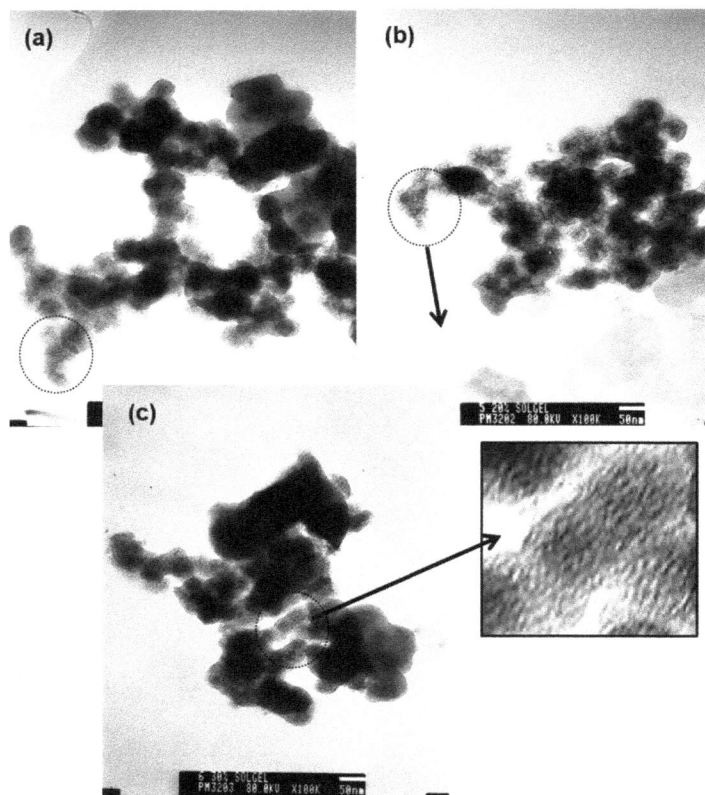

Figure 8. TEM micrographs corresponding to Nb-MCM-41 mesoporous solids obtained using the sol-gel method with niobium percentages of (**a**) 10%, (**b**) 20%, and (**c**) 30%.

3.2. Adsorption Kinetics and Results of Photodegradation Effect

The methylene blue adsorption kinetics corresponding to mesoporous solids are shown in Figure 9. These curves show marked differences in dye adsorption rates obtained by the solids, depending on the methodology used in the incorporation of niobium, as well as on the percentage of this cation on the mesoporous silica MCM-41. The solids with the highest dye adsorption capacity were those containing 10% Nb, obtained through the sol-gel method, and 20% Nb obtained using the incipient impregnation method. This result indicates that greater uniformity of niobium distribution on the surface of the mesoporous solid guarantees higher adsorption efficiency, as indicated by the Raman spectra (Figure 2).

Further, solids with higher adsorption capacity did not show leaching or formation of Nb_2O_5 clusters. What is evident in the adsorption kinetics (Figure 9) is that the Nb percentage does not notably affect dye adsorption.

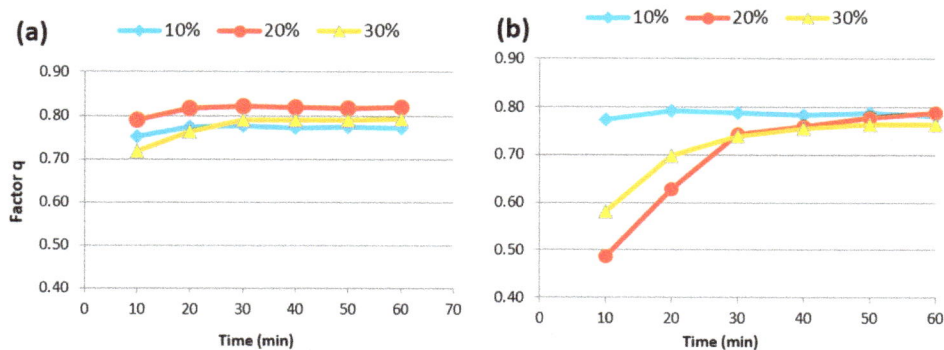

Figure 9. Adsorption kinetics of Nb-MCM-41 mesoporous solids obtained using the sol-gel (**a**) and impregnation method (**b**) (Factor q: mg methylene blue adsorbed/mg photocatalyst).

Although all the solids exhibit nearly the same total adsorption capacity, the behavior of the curves, as shown in Figure 9, is different, leading to the conclusion that the adsorption mechanisms were different. This could be due, in the main, to the differences in the microstructure of the samples, since as seen in the XRD at low angles, they clearly show a different mesophase, which could be closely related to the diffusion of the reagents. It is necessary, in future, to carry out a careful study to learn more about the mechanisms of adsorption as regards these samples and relate them to their characteristics (Table 1).

The process of photodegradation of methylene blue brought about by the Nb/MCM-41 mesoporous solids synthesized in this work is illustrated in Figures 10 and 11. The results show that for both methods of Nb incorporation, solids with 20% Nb showed a better photodegradation response. By comparing the curves, it is evident that the solid obtained using the sol-gel method showed a much higher degradation percentage than the solid obtained using incipient impregnation, with comparative figures of 60.6% and 16.7%, respectively (Figures 10 and 11).

Additionally, considering these results, and taking into account those of the Raman spectroscopy (Figure 2), it can be concluded that better niobium dispersion on the silica leads to a reduced photodegradation effect. In other words, although the uniform incorporation of niobium on mesoporous silica favored adsorption capacity, this condition notably diminished the photodegradation effect. This is not the case in the samples containing 10% Nb, where the reverse is true, *i.e.*, better niobium dispersion over the mesoporous silica indicates a greater photodegradation effect. This shows that it is necessary to conduct more systematic studies on both the niobium percentages in the sample and the uniformity of niobium dispersion, in the context of the effect of these on photodegradation capacity. Despite the fact that the 30% Nb-MCM-41 sample contains more niobium, as was determined in SEM-EDS, it may be that some of the Nb is not available to take part in photocatalysis, for example the Nb can be deposited in the zeolite porosities. Furthermore, as the amount of niobium in photocatalysts is increased, the surface area decreases, which indicates that there are fewer active sites available.

What is indeed evident in the results obtained is that the solids synthesized using the sol-gel method presented a greater photodegradation response than those obtained using the incipient impregnation method (Figures 10 and 11). This behavior could be justified considering that the pores observed in the TEM micrographs (Figure 8) of solids from the sol-gel method are more uniform than those observed in the incipient impregnation samples. This characteristic would provide a bigger surface area in the former samples, improving the photocatalytic process.

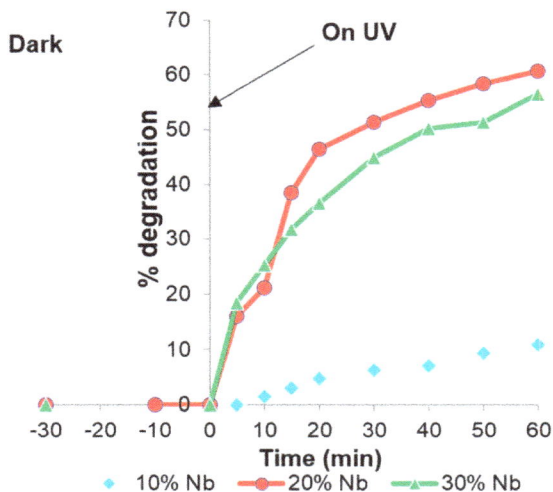

Figure 10. Degradation percentage of methylene blue, in function of irradiation time (sol-gel method).

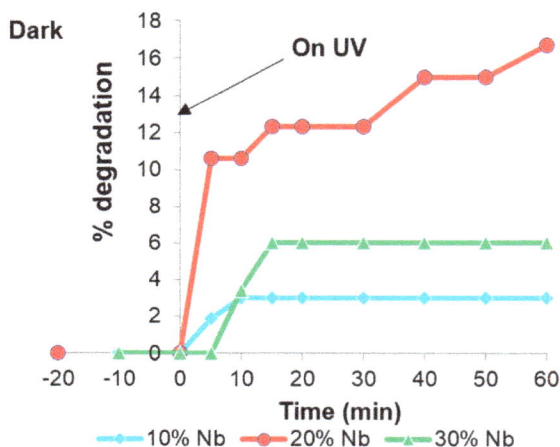

Figure 11. Degradation percentage of methylene blue *vs.* irradiation time (incipient impregnation method).

The photocatalysts studied in this work can be considered, mainly those obtained by impregnation as a result of their behavior and the results obtained, as another type of photocatalyst, *i.e.*, "single-site photocatalyst". In general, this single-site photocatalysis contains isolated polyhedral coordination transition metal oxides such as the oxides of Ti^{4+}, V^{5+}, Cr^{6+} and Mo^{6+} as active sites [30]; in this study, Nb^{5+} oxide. These species were designed on a mesoporous molecular MCM-41 surface and were considered to be highly dispersed at the atomic level on its surface (mainly in the samples obtained by impregnation) or within the frameworks and to be well-defined active sites. These local structures of our single-site Nb photocatalysts could explain the results obtained using Raman spectroscopy (Figure 2) and the behaviour of photocatalysts in the reduction of methylene blue (Figures 10 and 11). The photocatalytic properties of single-site photocatalysts Nb-MCM-41 could be attributed to the ligand to metal charge transfer (LMCT) process of the isolated niobium oxide with tetrahedral coordination. Under light irradiation, the charge transfer excited triplet state of niobium oxide was formed and electron-hole pairs were localized in close proximity to its photo-excited states [30]. Single-site photocatalysts exhibit unique and fascinating photocatalytic performances based on the reactions of specific photo-excited species. This process cannot be carried out on bulk photocatalysts [30]. Finally, the method of synthesis is clearly important in obtaining the single-site photocatalyst Nb-MCM-41 and for its functionality.

4. Conclusions

Regarding Nb-MCM-41 synthesis, the solids synthesized using incipient impregnation show a hexagonally ordered MCM-41 type mesoporous structure, as observed in the X-ray diffractograms. Solids obtained by the sol-gel method, meanwhile, did not present this type of structure. However, the mesoporosity of the systems is evident in the transmission electron micrographs that indicate a mesoporous-type ordered structure, although the family to which they correspond would still need to be determined.

Using incipient impregnation, better dispersion was observed on the silica, given that for up to 20% of incorporation, no Nb_2O_5 leaching was observed. This method thus shows greater advantages for niobium incorporation in silica. As such, greater dispersion appears to generate a less homogeneous pore size and a reduced photodegradation effect.

The Nb-MCM-41 mesoporous solids synthesized using the sol-gel method gave a greater photodegradation response than those synthesized through the incipient impregnation method. The highest effectiveness in photodegradation of the methylene blue molecule was obtained with the mesoporous solid synthesized through the sol-gel method, at a level of 20% Nb. This suggests that it is important to determine the structure obtained through the sol-gel method, to understand better the interaction between silica and niobium and the resulting effect on photodegradation. The results obtained in this study reiterate the importance of the method of synthesis used to obtain the mesoporous materials on their functionality.

Acknowledgements

The authors would like to thank CBMM for providing the niobium metal precursor. We are very grateful to the University of Cauca for offering the use of its laboratory facilities and making the necessary research time available. We are also grateful to Colin McLachlan for suggestions relating to the English text. This work was supported by project ID 4032—VRI of the University of Cauca.

Author Contributions

C.D.G and J.E.R both coordinated and supervised the different research projects involving synthesis and photocatalytic applications; C.D.G prepared the manuscript. All authors read and approved the manuscript.

Conflicts of Interest

The authors declare no conflict of interest.

References

1. Gaya, U.I. *Heterogeneous Photocatalysis Using Inorganic Semiconductor Solids*; Springer Science + Business Media: Dordrecht, The Netherlands, 2014.
2. Kaneko, M.; Okura, I. *Photocatalysis: Science and Technology*; Kodansha–Springer: New York, NY, USA, 2002.
3. Hoffmann, M.R.; Martin, S.T.; Choi, W.; Bahnemann, D.W. Environmental applications of semiconductor photocatalysis. *Chem. Rev.* **1995**, *95*, 69–96.
4. Coronado, J.M.; Fresno, F.; Hernández-Alonso, M.D.; Porteña, R. *Design of Advanced Photocatalytic Materials for Energy and Environmental Applications*; Springer-Verlag: London, UK, 2013.
5. Zhou, B.; Raja, R.; Han, S.; Somorjai, G.A. *Nanotechnology in Catalysis*; Springer: New York, NY, USA, 2007; Volume 3.
6. Brus, L. Electronic wave functions in semiconductor clusters: experiment and theory. *J. Phys. Chem.* **1986**, *90*, 2555–2560.
7. Nozik, A.J.; Williams, F.; Nenadovic, M.T.; Rajh, T.; Micic, O.I. Size quantization in small semiconductor particles. *J. Phys. Chem.* **1985**, *89*, 397–399.
8. Stroyuk, O.L.; Kuchmiy, S.Y.; Kryukov, A.I.; Pokhodenko, V.D. *Semiconductor Catalysis and Photocatalysis on the Nanoscale*; Nova Science Publishers, Inc.: New York, NY, USA, 2010.
9. Shchukin, D.G.; Sviridov, D.V. Photocatalytic processes in spatially confined micro- and nanoreactors. *J. Photochem. Photobiol. C* **2006**, *7*, 23–39.
10. Chen, H.; Chen, S.; Quan, X.; Zhang, Y. Structuring a TiO_2-based photonic crystal photocatalyst with Schottky junction for efficient photocatalysis. *Environ. Sci. Technol.* **2010**, *44*, 451–455.
11. Kim, S.; Kwak, S.Y. Photocatalytic inactivation of *E. coli* with mesoporous TiO_2 coated film using the film adhesion method. *Environ. Sci. Technol.* **2009**, *43*, 148–151.
12. Yang, H.-C.; Lin, H.-Y.; Chien, Y.-S.; Wu, J.C.-S. Mesoporous TiO_2/SBA-15 and Cu/TiO_2/SBA-15 composite photocatalysts for photoreduction of CO_2 to methanol. *Cat. Lett.* **2009**, *131*, 381–387.
13. Li, X.; Lu, K.; Deng, K.; Tang, J.; Su, R.; Sun, J.; Chen, L. Synthesis and characterization of ZnO and TiO_2 hollow spheres with enhanced photoreactivity. *Mat. Sci. Eng. B* **2009**, *158*, 40–47.

14. Ji, P.; Zhang, J.; Chen, F.; Anpo, M. Ordered mesoporous CeO_2 synthesized by nanocasting from cubic Ia3d mesoporous MCM-48 silica: Formation, characterization and photocatalytic activity. *J. Phys. Chem. C* **2008**, *112*, 17809–17813.

15. Chen, Y.; Hu, L.; Wang, M.; Min, Y.; Zhang, Y. Self-assembled Co_3O_4 porous nanostructures and their purification and their photocatalytic activity. *Colloids Surf. A* **2009**, *336*, 64–68.

16. Ziolek, M.; Nowak, I. Synthesis and characterization of niobium-containing MCM-41. *Zeolites* **1997**, *18*, 356–360.

17. Ziolek, M.; Sobezak, I.; Nowak, I.; Decyk, P.; Lewandowaska, A.; Kujawa, J. Nb-containing mesoporous molecular sieves—Possible application in the catalytic processes. *Micro. Mesop. Mater.* **2000**, *35–36*, 195–207.

18. Ziolek, M.; Sobezak, I.; Lewandowska, A.; Nowak, I.; Decyk, P.; Renn, M.; Jankowska, B. Oxidative properties of niobium-containing mesoporous silica catalysts. *Catal. Today* **2001**, *70*, 169–181.

19. Chen, X.; Yu, T.; Fan, X.; Zhang, H.; Li, Z.; Ye, J.; Zou, Z. Enhanced activity of mesoporous Nb_2O_5 for photocatalytic hydrogen production. *Appl. Surf. Sci.* **2007**, *253*, 8500–8506.

20. Lin, H.-Y.; Yang, H.-C.; Wang, W.-L. Synthesis of mesoporous Nb_2O_5 photocatalysts with Pt, Au, Cu and NiO cocatalyst for water splitting. *Catal. Today* **2011**, *174*, 106–113.

21. Lopes, O.F.; Paris, E.C.; Ribeiro, C. Synthesis of nanoparticles through the oxidant peroxide method applied to organic pollutant photodegradation: A mechanistic study. *Appl. Catal. B Environ.* **2014**, *144*, 800–808.

22. Oliveira, L.C.A.; Ramalho, T.C.; Goncalves, M.; Cereda, F.; Carvalho, K.T.; Nazzarro, M.S.; Sapag, K. Pure niobia as catalyst for the oxidation of organic contaminants: mechanism study via ESI-MS and theoretical calculations. *Chem. Phys. Lett.* **2007**, *446*, 133–137.

23. Esteves, A.; Oliveira, L.C.A.; Ramalho, T.C.; Goncalves, M.; Anastacio, A.S.; Carvalho, H.W.P. New materials base on modified synthetic Nb_2O_5 as photocatalyst for oxidation of organic contaminants. *Catal. Commun.* **2008**, *10*, 330–332.

24. Gallo, J.M.R.; Paulino, I.S.; Schuchardt, Ulf. Cyclooctene epoxidation using Nb-MCM-41 and Ti-MCM-41 synthesized at room temperature. *Appl. Catal. A* **2004**, *266*, 223–227.

25. Gallo, J.M.R.; Pastore, H.O.; Schuchardt, U. Silylation of [Nb]-MCM-41 as an efficient tool to improve epoxidation activity and selectivity. *J. Catal.* **2006**, *243*, 57–63.

26. Nowak, I.; Misiewicz, M.; Ziolek, M.; Kubacka, A.; Corte´s Corberán, V.; Sulikowski, B. Catalytic properties of niobium and gallium oxide systems supported on MCM-41 type materials. *Appl. Catal. A* **2007**, *325*, 328–335.

27. Anilkumar, M.; Hölderich, W.F. Highly active and selective Nb modified MCM-41 catalysts for Beckmann rearrangement of cyclohexanone oxime to ε-caprolactam. *J. Catal.* **2008**, *260*, 17–29.

28. An, X.; Gao, C. Synthesis of mesoporous N-doped TiO_2/ZnAl-layered double oxides nanocomposite for efficient photodegradation of methyl orange. *Mater. Sci. Semicond. Process.* **2015**, *34*, 162–169.

29. Solmaza, A.; TimurDogu, S.B. Synthesis and characterization of V, Mo and Nb incorporated micro–mesoporous MCM-41 materials. *Mater. Chem. Phys.* **2011**, *125*, 148–155.
30. Kamegawa, T.; Yamashita, H. Solar energy conversion using single-site photocatalysts. In *New and Future Developments in Catalysis: Solar Photocatalysis*; Suib, L., Ed.; Elsevier: Amsterdan, The Netherlands, 2013; pp. 103–119.

Evaluation of Photocatalytic Properties of Portland Cement Blended with Titanium Oxynitride (TiO$_{2-x}$N$_y$) Nanoparticles

Juan D. Cohen, G. Sierra-Gallego and Jorge I. Tobón

Abstract: Photocatalytic activity of Portland cement pastes blended with nanoparticles of titanium oxynitride (TiO$_{2-x}$N$_y$) was studied. Samples with different percentages of TiO$_{2-x}$N$_y$ (0.0%, 0.5%, 1%, 3%) and TiO$_2$ (1%, 3%) were evaluated in order to study their self-cleaning properties. The presence of nitrogen in the tetragonal structure of TiO$_2$ was evidenced by X-ray diffraction (XRD) as a shift of the peaks in the 2θ axis. The samples were prepared with a water/cement ratio of 0.5 and a concentration of Rhodamine B of 0.5 g/L. After 65 h of curing time, the samples were irradiated with UV lamps to evaluate the reduction of the pigment. The color analysis was carried out using a Spectrometer UV/Vis measuring the coordinates CIE (Commission Internationale de l'Eclairage) L^*, a^*, b^*, and with special attention to the reddish tones (Rhodamine B color) which correspond to a^* values greater than zero. Additionally, samples with 0.5%, 1%, 3% of TiO$_{2-x}$N$_y$ and 1%, 3% of TiO$_2$ were evaluated under visible light with the purpose of determining the Rhodamine B abatement to wavelengths greater than 400 nm. The results have shown a similar behavior for both additions under UV light irradiation, with 3% being the addition with the highest photocatalytic efficiency obtained. However, TiO$_{2-x}$N$_y$ showed activity under irradiation with visible light, unlike TiO$_2$, which can only be activated under UV light.

Reprinted from *Coatings*. Cite as: Cohen, J.D.; Sierra-Gallego, G.; Tobón, J.I. Evaluation of Photocatalytic Properties of Portland Cement Blended with Titanium Oxynitride (TiO$_{2-x}$N$_y$) Nanoparticles. *Coatings* **2015**, *5*, 465-476.

1. Introduction

Nanotechnology has become a very important research topic in recent years for the scope and variety of applications in almost all fields of engineering. Mainly though, it has been important in the development of new materials. For example, nanotechnology has been implemented in the study of building materials in order to improve the mechanical properties and give the materials other additional properties. In these studies, the addition of carbon nanotubes to reduce fissures and improve mechanical properties [1], as well as the insertion of nanoparticles of silica (n-SiO$_2$) [2] and iron oxide (Fe$_2$O$_3$) for the sealing of nanopores [3] has been highlighted. Other types of features have been given to cement, when broadband semiconductor nanoparticles are added, such as the decontaminating of air, and the self-cleaning in buildings, through photocatalytic activity [4]. These properties are obtained, for example, when photocatalytic reactions are performed in the nanoparticles interface incorporated in the cementitious matrices. These reactions are defined as electrochemical processes by absorption of radiant energy (UV light) within the photocatalyst, which is usually a semiconductor with broadband [5]. When photons of a required wavelength (photons with energy higher than the band gap of the photocatalyst), are absorbed by the photocatalyst, electrons from the valence band are promoted to the conduction band crossing through the band gap (forbidden region for electronic states), and electron-hole pairs are generated. These pairings carry opposed free charges in the absence of an electric field, and recombine rapidly (in approximately of 30 ns), releasing excess energy, mainly as heat [6]. If there are previously adsorbed chemical species in the catalyst surface, recombination is prevented and redox reactions occur between these species and the photogenerated pairs. In this case, the electrons (−) react with oxidizing agents, and the holes (+) react with reducing agents. Photocatalysis is normally effected under aerobic conditions, (when oxygen acts as an acceptor species) and will react with the electrons (e^-) to form a superoxide radicals ($O_2{}^{•-}$). In turn, the water is used as a reducing agent, and it will react with the holes (h^+) to form hydroxyl radicals ($OH^•$) [5]. The following Equations (1)–(3) correspond to the reactions carried out in the photocatalyst interface:

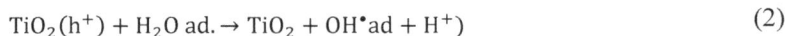

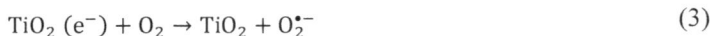

$$TiO_2 + hv \rightarrow TiO_2(e^- + h^+) \tag{1}$$

$$TiO_2(h^+) + H_2O \text{ ad.} \rightarrow TiO_2 + OH^•\text{ad} + H^+) \tag{2}$$

$$TiO_2(e^-) + O_2 \rightarrow TiO_2 + O_2^{•-} \tag{3}$$

Studies with photocatalytic cements have been performed mostly by adding TiO_2 nanoparticles, and UV radiation as a photon source. The use of TiO_2 is due to its low toxicity and high stability compared to other semiconductors studied for this application (ZnO, CdS, WO_3) [6]. The NO_x abatement has been evaluated for cement pastes added with TiO_2 nanoparticles at different ratios of rutile and anatase [7]. In this same way, there has been reported the mineralization of different VOCs (Volatile Organic Compounds) by means of building materials added with TiO_2 [8]. The self-cleaning ability in these kind of cements is evaluated by standard UNI 11259, using Rhodamine B as an organic dye. In this case, important results have been obtained with mortars and cement pastes, using TiO_2 as addition [9–11]. The Rhodamine B is selected for evaluating this property in cements, mainly because it is very soluble in water, its discoloration can be followed by colorimetry and it has little sensitivity to the alkalinity of cementitious materials [10]. Additionally, the rhodamine B has polycyclic aromatic hydrocarbons in its chemical structure, compounds that are usually found as pollutants in urban environments. On the other hand, it is known that the main weakness of these cements is given by the nature of TiO_2, which is only activated in the presence of UV light. The wavelengths in the UV region of the solar spectrum correspond to a low percentage and, thus, there is minimal use of stimulus of light when implementing these cements into a real life scenario. An alternative though, according to the research performed by Asahi, is to modify the photocatalyst with impurities to improve the efficiency of the process. This research found that the presence of nitrogen in the structure of TiO_2 as substitution by oxygen creates Ti–N bonds and electronic states in forbidden energy levels [12,13]. This anionic partial substitution increases the range of the solar spectrum absorbed [14]. In this case, nanoparticles of TiO_xN_y can absorb wavelengths between 365 nm and 500 nm [15,16]. An approach of the behavior of these nanoparticles in building materials was done by mixing nano-$TiO_{2-x}N_y$ with calcium carbonate ($CaCO_3$). This blend reduced the concentration of NO_x and 2-propanol between 360 and 436 nm [17]. The addition of nano-$TiO_{2-x}N_y$ to the cement is entirely unknown; therefore, the main objective of this research is to discuss the behavior of these nanoparticles in the cement and obtain a construction material with a self-cleaning property in the presence of UV and visible light.

2. Experimental Section

2.1. Raw Materials

Samples were prepared using commercial Portland white cement type I from the Colombian company Cementos Argos. The Titanium Oxynitride ($TiO_{2-x}N_y$) and Titanium Dioxide (TiO_2), commercial references were provided by Chinese NaBond Technologies Co. (Shenzhen, China) and German companies Evonik-Degussa (Essen, Germany), respectively. In turn, according to the manufacturer, the $TiO_{2-x}N_y$ has 22% Nitrogen in its structure. The Rhodamine B was provided by the Colombian enterprise Codim. Table 1 shows the specifications of each nanoparticle and the crystalline phases and semi-quantitative estimations using X-ray diffraction (XRD, Phillips X'pert).

Table 1. Characteristics of nanoparticles added.

Sample	BET* (m^2/g)	Dv.50* (nm)	Crystalline phases (%) **
P25 (TiO_2)	50 ± 15	21	Anatase 87 Rutile 13
$TiO_{2-x}N_y$	31	30	Anatase 84 Rutile 16

*According to the manufacturer's specifications; ** Semi-quantitative estimation by XRD.

2.2. Preparation and Evaluation of the Samples

Samples were prepared into disc-shaped specimens of white cement paste with a 1.62 cm diameter and a 0.3 cm thickness. Deionized water was used for a water/cement ratio of 0.5. Previously, the nanoparticles were added in the mixed water and dispersed using a superplasticizer (Plastol HR-DF), provided by the company Toxement in Colombia. The dispersant agent was used with 19% by weight relative to the nanoparticles. Afterwards, this suspension was sonicated with 90 W and 35 kHz frequency (VWR® Symphony™ 97043-992) for 15 min [18]. Samples were prepared with three different percentages of cement content, for $TiO_{2-x}N_y$ 0.5%, 1%, 3%. Other samples were prepared with two levels of TiO_2, 1% and 3%. After, the samples were cured in a wet room for 65 h. Additionally, samples were prepared with 0.5%, 1%, 3% of $TiO_{2-x}N_y$, and another set with 1% and 3% of TiO_2, in order to evaluate the Rhodamine B abatement in visible light. Following the provisions of the standard [19], the pastes were immersed in water with Rhodamine B for one hour, at a concentration of 0.5 g/L in order to ensure a uniform color for each sample. The samples were irradiated with UV light using lamps emitting a wavelength spectrum between 350 and 400 nm (Phillips, Amsterdam, The Netherlands, Actinic BL TL-D(K), 30 W). In order to evaluate the behavior of the samples in visible light, a fluorescent lamp was used with a wavelength spectrum between 410 and 560 nm (Lite-Way, C13101, 24 W). The samples were irradiated to 20 W/m^2 ± 0.3 for the UV experimental setup, as set out in the standard [19]. For the setup with visible light, the samples were irradiated to 10 W/m^2 ± 0.5. For this setup, the height between the lamp and the samples was determined using a radiometer (PMA 2210, Solar Light Co., Philadelphia, PA, USA) coupled to a probe to detect wavelengths between 400 and 700 nm (LDC 21300, Solar Light Co., Philadelphia, PA, USA). To determine the color coordinates CIE (Commission Internationale de l'Eclairage)

L*a*b* [20], a UV/Vis Spectrometer was used (BWTEK Inc., Newark, NJ, USA, GlacierTM X, BTC112E), coupled with a probe for measuring reflection and backscatter (Ocean Optics, QR200-7-UV-VIS) and a support to hold it at 45° angle to the samples (Ocean Optics, Dunedin, FL, USA, RPH-1).

3. Results and Discussion

3.1. X-ray Diffraction (XRD)

Figure 1 corresponds to diffractograms of $TiO_{2-x}N$ and TiO_2, highlighting anatase and rutile phases. The shifting of peaks to the left is evidence of an enlargement of the unit cell, demonstrating the insertion of nitrogen atoms into the crystalline structure. Since the TiO_2 usually has oxygen vacancies that introduce localized states of Ti^{+3} [6], it is possible to insert some impurities that are compatible and dope the catalyst to improve performance. In this case, the nitrogen (N^{-3}) is likely to enter into the structure for having an atomic radius and electronegativity similar to O^{-2}. In this manner, the nitrogen supplies the anionic deficiency. Precisely, it is the nitrogen inserted in the interstices as a substitution of oxygen, which induces occupied electronic states on intervals known as the forbidden energy band (band gap) [13]. This process can facilitate the formation of electron-hole pairs, using stimuli such as electromagnetic radiation of wavelengths greater than 400 nm.

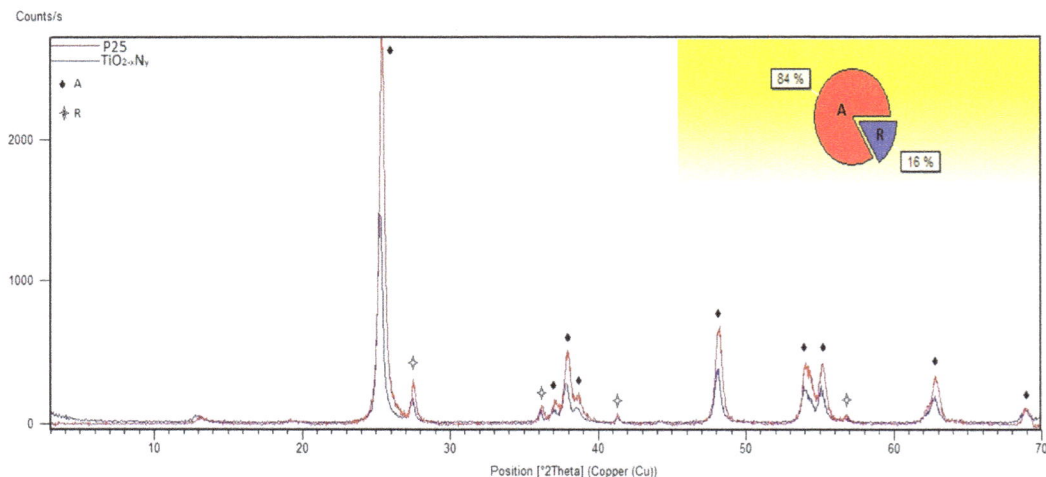

Figure 1. X-ray diffractogram (XRD) of Aeroxide P25 (TiO_2) *vs.* $TiO_{2-x}N_y$; phases anatase (A) and rutile (R).

3.2. Transmission Electron Microscope (TEM)

TEM images of nanoparticles of TiO_2 and $TiO_{2-x}N_y$, are reported in Figure 2. Both have almost the same distribution of phases, according to the diffractogram XRD viewed above. The anatase with a hexagonally shaped rounded, has a size about 20 nm (Figure 2a), while rutile shows a hexagonally shape and a size about 60 nm for TiO₂ (Figure 2a). In Figure 2b, $TiO_{2-x}N_y$ nanoparticles have an overall size of about 30 nm.

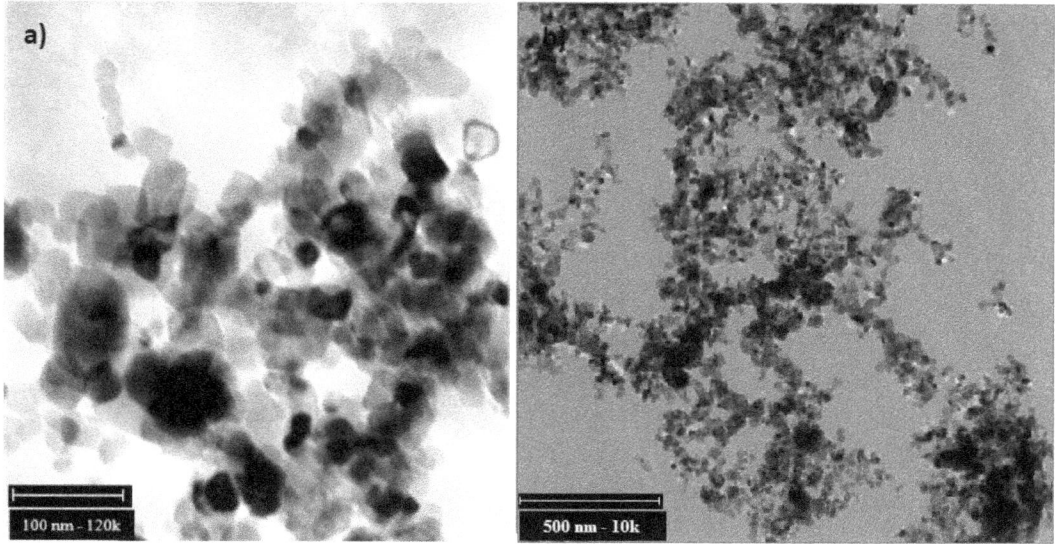

Figure 2. TEM images for nanopowders: (**a**) Aeroxide P25 (TiO₂) [11]; (**b**) $TiO_{2-x}N_y$ (NaBond Technologies Co., Shenzhen, China).

3.3. Abatement of Rhodamine B

The Rhodamine B degradation mechanism passes through two different pathways; the first being the deethylation process and the second being the destruction of the chromophore structure [21]. Such a dual mechanism seems to depend on the nature of the light in the photoinduction. Under UV-Vis conditions, there is no selectivity and the destruction of the chromophore is also involved during the degradation. [22]. Additionally, the results obtained during discoloration of the samples are observed in Figure 3, where Δa^* is defined as the change in coordinate a^* during the exposure time of the samples to UV light; being the initial hue at $t = 0$ and the final coordinate of the same sample at time t as shown below:

$$\Delta a^* = a_0^* - a_t^* \tag{4}$$

In this case, the coordinate a^* corresponds to a reddish hue for positive values, according the color-order system specified by the Commission Internationale de l'Eclairage [20]. The highest photoactivity was observed for samples with a 3% addition of $TiO_{2-x}N_y$ and TiO₂, as similarly reported by other works [10,11]. The control samples were also evaluated, to determine the catalytic reactions by photolysis or by temperature.

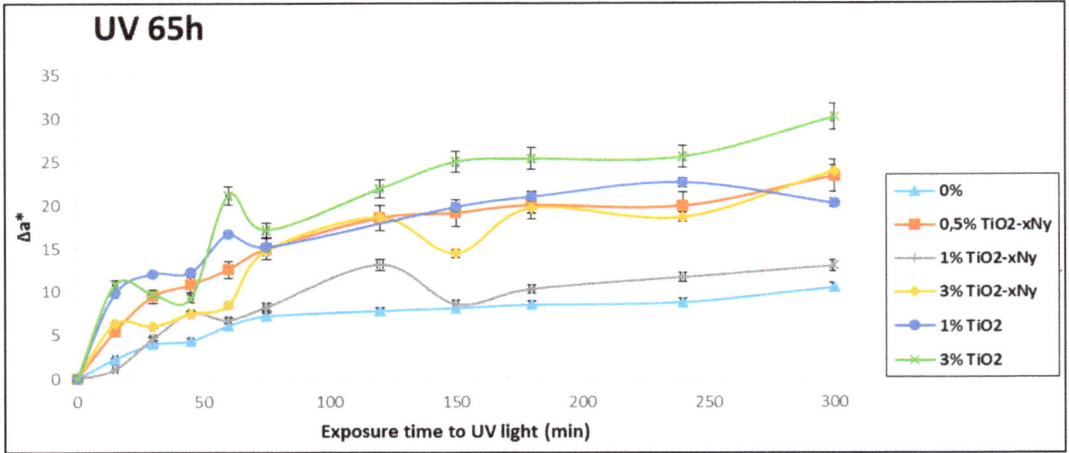

Figure 3. Change in Δa^* during five hour exposure to UV light with different addition percentages of $TiO_{2-x}N_y$ and TiO_2 for samples at 65 h of curing age.

There is a difference between the specimens with 3% of TiO_2 facing the other samples was observed, and the values of 0.5% of $TiO_{2-x}N_y$ were similar to the samples with 1% of TiO_2. It is noted that 0.5% of $TiO_{2-x}N_y$ showed a more significant change than 1% of $TiO_{2-x}N_y$; this behavior can be explained in two ways. The first explanation is that the action of the hydrated products forms a layer around nano- $TiO_{2-x}N_y$, which protects nanoparticles from abrasion and may decrease the photocatalytic activity [23]. Here, the nanoparticles act as nucleation sites in cement hydration increasing the accumulation of hydration products [24]. The second way is that the decrease in photocatalytic activity is explained by re-agglomeration of nanoparticles in the matrix of the cement as a consequence of low dispersion. This occurs also in the samples with 3% of $TiO_{2-x}N_y$, but the generation of active sites seems to be a more important factor in this case; therefore, the abatement of Rhodamine B is improved. Even with this information, it is necessary to find other evidence, such as the generation of hydration products and the relationship of them with the content of nanoparticles in the cement and the photoactivity, in future works in order to ratify this hypothesis.

In order to determine a percentage of decrease of color, the photocatalytic efficiency coefficient (ε) was calculated and is defined as:

$$\varepsilon = \frac{A(a_0^*) - A(a^*)}{A(a_0^*)} \times 100 \tag{5}$$

where $A(a_0^*)$ corresponds to an ideal state in which the initial coordinate remains constant during t_f time, as established by Equation (6), and $A(a^*)$ the area under the real curve of a^*, as observed in Equation (7) [10].

$$A(a_0^*) = t_f \times a_0^* \tag{6}$$

$$A(a^*) = \int_{t_0}^{t_f} a^* dt \tag{7}$$

Figure 4 shows the photocatalytic efficiency coefficients of the samples (cured 65 h) after 5 h of UV irradiation. Coefficients greater than those obtained for control samples (0%), are considered photocatalytic reactions and are products of the interaction between photons and the nanoparticles of photocatalyst. Otherwise, these coefficients are considered as other kinds of interactions, such as photolysis or thermolysis. It is noted, that all the additions showed photocatalytic activity and efficiency above the standard sample. In addition, a similar coefficient between the samples with 0.5% of $TiO_{2-x}N_y$ and 1% $TiO_{2-x}N_y$ was observed, different to results obtained in another research [11]. As explained above, this could be due to the re-aglomeration effect of the nanoparticles in the matrix cementitious. Similarly, a uniform performance of TiO_2 3% was observed for $TiO_{2-x}N_y$ 3%.

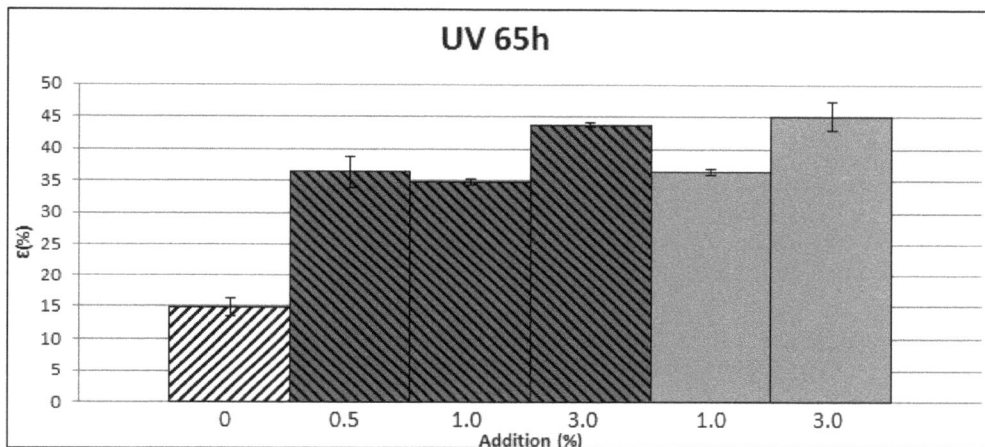

Figure 4. Photocatalytic efficiency Coefficient (ε) for samples in UV light with different addition percentages of $TiO_{2-x}N_y$ and TiO_2 for samples at 65 h of curing age, ▨ control sample, ▧ $TiO_{2-x}N_y$, ☐ TiO_2. Error bars are standard deviation.

Some samples were evaluated as well in visible light. In Figure 5, the comparison between samples of $TiO_{2-x}N_y$ with 0.5%, 1%, 3% and TiO_2 with 1%, 3% for the change in Δa^* *versus* exposure time is shown. It is noted that TiO_2 did not show photoactivity in visible light, mainly because of its wide band gap that only lets the electrons jump to the conduction band by stimuli of wavelengths below 400 nm.

Additionally, the photocatalytic efficiency coefficient for these samples was calculated. In Figure 6, the comparison between coefficients for samples with 0.5%, 1%, 3% of $TiO_{2-x}N_y$, and 1%, 3% of TiO_2, including control samples are reported. TiO_2 samples obtained values below the samples with 0%. The minimal change in a^* is attributed to other mechanisms, not to photocatalytic reactions. When percentages equal or below are obtained in comparison with the control sample (0%), there is no evidence of photocatalytic interactions; therefore, the change in hue is attributed to photolysis and thermolysis reactions. It is possible that TiO_2 nanoparticles create a shielding effect in the samples, blocking photonic absorption in the pigment and preventing photolysis reactions.

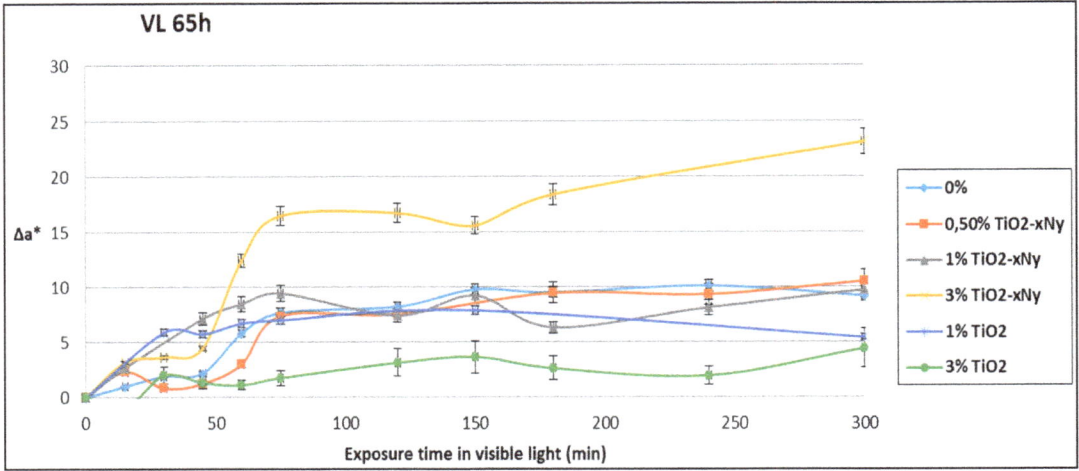

Figure 5. Change in $\Delta a*$ during 5 h exposure to visible light with different addition percentages of $TiO_{2-x}N_y$ and TiO_2 for samples at 65 h of curing age.

Table 2. Standard deviation of samples with $TiO_{2-x}N_y$ in visible light.

Sample	% $TiO_{2-x}N_y$	L_0*	a_0*	b_0*	$L*$	$a*$	$b*$	ΔE	Variance $a*$	Standard Deviation $a*$
A_1	0	64.60	67.90	−35.17	80.33	58.71	−23.45	21.65	0.000648	0.025455844
A_2	0	68.84	69.94	−34.39	79.30	58.75	−22.75	19.24		
B_1	0.5	66.15	68.09	−35.43	81.79	57.55	−23.25	22.47	5.5677845	2.359615329
B_2	0.5	63.429	67.77	−33.7	79.59	54.213	−23.807	23.3		
B_1	1	67.405	49.809	−32.181	82.102	40.085	−14.098	25.24	3.1777205	1.782616195
B_2	1	69.915	50.46	−34.871	81.559	42.606	−16.775	22.905		
C_1	3	75.51	63.207	−31.92	83.163	40.038	−14.787	29.81	0.2628125	0.512652416
C_2	3	74.202	60.164	−32.178	77.017	40.763	−15.653	25.63		

Table 3. Standard deviation of samples with TiO_2 in visible light.

Sample	% TiO_2	L_0*	a_0*	b_0*	$L*$	$a*$	$b*$	ΔE	Variance $a*$	Standard Deviation $a*$
D_1	1	77.189	56.42	−33,115	88.339	50.988	−21.831	16.76	0.006728	0.082024387
D_2	1	74.513	57.788	−33,322	64.162	51.104	−21.06	17.38		
E_1	3	71.784	63.023	−37,562	70.132	58.578	−26.049	12.45	8.602952	2.933078928
E_2	3	73.469	63.501	−37,292	79.723	62.726	−27.142	11.94		

The tables above show some data of the samples evaluated in visible light. Standard deviation was calculated for $a*$ coordinate in all the samples. At the same time, the ΔE was calculated. This represents the change in color that is defined as the difference between two points of values, in the Euclidean sense [20]. The higher the ΔE, the higher the change in color. ΔE is defined as:

$$\Delta E = \sqrt{(L_0^* - L_t^*)^2 + (a_0^* - a_t^*)^2 + (b_0^* - b_t^*)^2} \tag{8}$$

It is noteworthy that the band gap of TiO_2 corresponds to 3.2 eV per atom; therefore, to do the electronic jump from the valence band to the conduction band and generate redox species, it is necessary to induce the particles and a suitable stimulus in terms of light or energy photons located in the range of near ultraviolet (UVA). On the other side, the inclusion of nitrogen atoms in the structure of TiO_2, as mentioned above, causes a decrease in the band gap, achieving the previously stated electronic jump more easily. In this form, it is possible to have photocatalytic activity in wavelengths between 400 nm and 700 nm. This can be observed in specimens with 3% of $TiO_{2-x}N_y$ that showed a percentage above the others. Samples with a low percentage of $TiO_{2-x}N_y$ showed low photoactivity, possibly due to the characteristics of the nanoparticles used. It is possible that commercial nano-$TiO_{2-x}N_y$ has a low doping degree of nitrogen and, for this reason, it requires high percentages in order to improve the efficiency. These results show that it is possible to obtain cements photoactivated with a self-cleaning property in visible light, but it is necessary to evaluate them in other conditions.

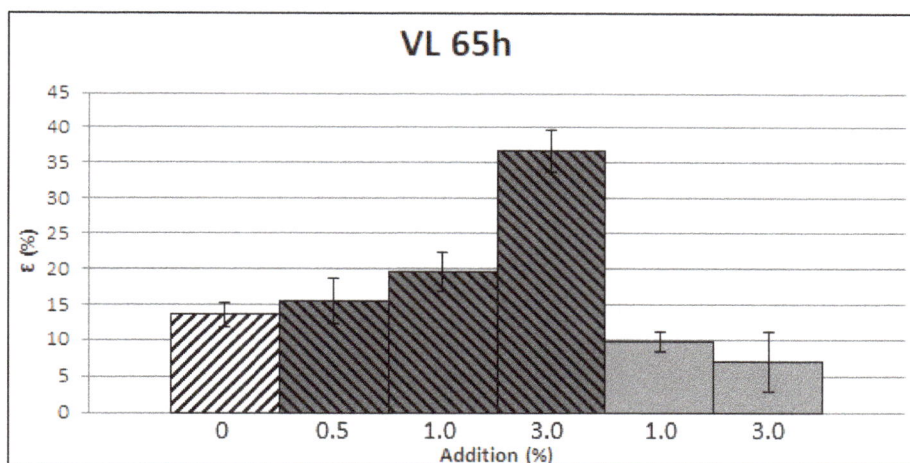

Figure 6. Photocatalytic efficiency Coefficient (ε) for samples in Visible light with 1%, 3% $TiO_{2-x}N_y$ and 1%, 3% TiO_2 for samples at 65 h of curing age, ▨ control sample, ▨ $TiO_{2-x}N_y$, ▨ TiO_2. Errors bars are standard deviation.

4. Conclusions

The addition of $TiO_{2-x}N_y$ nanoparticles to cement gives the ability to abate organic compounds present on the cementitious matrix, thus giving cement the attribute of being able to destroy contaminants by irradiating them with visible and UV light. With the best performance, samples with 3% $TiO_{2-x}N_y$ showed a similar behavior to 3% TiO_2 samples under UV irradiation. On the other hand, the results under visible light were different; the samples with TiO_2 did not show photocatalytic activity. In turn, these samples obtained a minimal abatement of dye by reactions of photolysis or thermolysis, and, possibly, a shielding effect was observed that prevented the photonic absorption in the pigment. In addition, the specimens with 3% of $TiO_{2-x}N_y$ obtained the highest photoactivity, and the samples with 1% of $TiO_{2-x}N_y$ showed a similar behavior to the control samples.

Cements with added $TiO_{2-x}N_y$ nanoparticles show potential to be applicable in the construction of photocatalytic structure, which can take advantage of much of the solar spectrum and artificial lighting for the degradation of organic pollutants present in these.

Acknowledgments

This work was supported by the "National program of projects for strengthening research, development and innovation in graduate 2014–2015" of the National University of Colombia. The authors would like to thank the university, but especially the Materials Characterization Laboratory of the faculty of mines from the National University of Colombia campus.

Author Contributions

Juan D. Cohen and Jorge I. Tobón have conceived and designed the experiments; Juan D. Cohen performed the experiments; G. Sierra-Gallego, Jorge I. Tobón and Juan D. Cohen analyzed the data. G. Sierra-Gallego and Jorge I. Tobón have contributed with the reagents and analysis tools. Juan D. Cohen wrote the paper.

Conflicts of Interest

The authors declare no conflict of interest.

References

1. Nochaiya, T.; Chaipanich, A. Behavior of multi-walled carbon nanotubes on the porosity and microstructure of cement-based materials. *Appl. Surf. Sci.* **2011**, *257*, 1941–1945.
2. Tobón, J.I.; Payá, J.J.; Borrachero, M.V.; Restrepo, O.J. Mineralogical evolution of Portland cement blended with sílica nanoparticles and its effect on mechanical strength. *Constr. Build. Mater.* **2012**, *36*, 736–742.
3. Oltulu, M.; Azahin, R. Single and combined effects of nano-SiO_2, nano-Al_2O_3 and nano-Fe_2O_3 powders on compressive strength and capillary permeability of cement mortar containing silica fume. *Mater. Sci. Eng. A* **2011**, *528*, 7012–7019.
4. Italcementi, S.P.A. Titanium Dioxide Based Photocatalytic Composites and Derived Products on a Metakaolin Support. U.S. Patent 8092586 B2, 10 January 2009.
5. Domènech, X.; Jardim, W.F.; Litter, M.I. Procesos avanzados de oxidación para la eliminación de contaminantes. In *Eliminación de contaminantes por fotocatálisis heterogénea*; Blesa, M.; Ed.; CYTED (Ibero-American Science and Technology for Sustainable Development): La Plata, Argentina, 2002; pp. 3–26.
6. Candal, R.; Bilmes, S.; Blesa, M. Semiconductores con actividad fotocatalítica. In *Eliminación de contaminantes por fotocatálisis heterogénea*; Blesa, M., Ed.; CYTED (Ibero-American Science and Technology for Sustainable Development): La Plata, Argentina 2002, pp. 79–101.

7. Cárdenas, C.; Tobón, J.; García, C.; Vila, J. Functionalized building materials: Photocatalytic abatement of NO$_x$ by cement pastes blended with TiO$_2$ nanoparticles. *Constr. Build. Mater.* **2012**, *36*, 820–825.

8. Strini, A.; Cassese, S.; Schiavi, L. Measurement of benzene, toluene, ethylbenzene and *o*-xylene gas phase photodegradation by titanium dioxide dispersed in cementitious materials using a mixed flow reactor. *Appl. Catal. B Environ.* **2005**, *61*, 90–97.

9. Diamanti, M.V.; Del Curto, B.; Ormellese, M.; Pedeferri, M.P. Photocatalytic and self-cleaning activity of colored mortars containing TiO$_2$. *Constr. Build. Mater.* **2013**, *46*, 167–174.

10. Ruot, B.; Plassais, A.; Olive, F.; Guillot, L.; Bonafous, L. TiO$_2$-containing cement pastes and mortars: Measurements of the photocatalytic efficiency using a rhodamine B-based colourimetric test. *Sol. Energy* **2009**, *83*, 1794–1801.

11. Cárdenas, C.; Tobón, J.; García, C. Photocatalytic properties evaluation of Portland white cement added with TiO$_2$-Nanoparticles. *Rev. LatinAm. Metal. Mat.* **2013**, *33*, 316–322.

12. Asahi, R.; Morikawa, T.; Ohwaki, T.; Aoki, K.; Taga, Y. Visible-light photocatalysis in nitrogen-doped titanium oxides. *Science* **2001**, *293*, 269–271.

13. Di Valentin, C.; Finazzi, E.; Pacchioni, G.; Selloni, A.; Livraghi, S.; Paganini, M.C.; Giamello, E. N-doped TiO$_2$: Theory and experiment. *Chem. Phys.* **2007**, *339*, 44–56.

14. Kitano, M.; Funatsu, K.; Matsuoka, M.; Ueshima, M.; Anpo, M. Preparation of nitrogen-substituted TiO$_2$ thin film photo catalysts by the radio frequency magnetron sputtering deposition method and their photo catalytic reactivity under visible light irradiation. *J. Phys. Chem. B* **2006**, *110*, 25266–25272.

15. Hong, Y.C.; Bang, C.U.; Shin, D.H.; Uhm, H.S. Band gap narrowing of TiO$_2$ by nitrogen doping in atmospheric microwave plasma. *Chem. Phys. Lett.* **2005**, *413*, 454–457.

16. Yin, S.; Aita, Y.; Komatsu, M.; Wang, J.; Tanga, Q.; Satoa, T. Synthesis of excellent visible-light responsive TiO$_{2-x}$N$_y$ photocatalyst by a homogeneous precipitation-solvothermal process. *J. Mater. Chem.* **2005**, *15*, 674–682.

17. Amadelli, R.; Samiolo, L.; Borsa, M.; Bellardita, M.; Palmisano, L. N-TiO$_2$ Photocatalysts highly active under visible irradiation for NO$_x$ abatement and 2-propanol oxidation. *Catal. Today* **2013**, *206*, 19–25.

18. Yousefi, A.; Allahverdi, A.; Hejazi, P. Effective dispersion of nano-TiO$_2$ powder for enhancement of photocatalytic properties in cement mixes. *Constr. Build. Mater.* **2013**, *41*, 224–230.

19. UNI 11259 Determinazione dell'attività fotocatalitica di leganti idraulici: Metodo della rodammina; UNI (Ente italiano di normazione): Milano, Italy, 2008.

20. Billmeyer, F.W.; Saltzman, M. *Principles of Color Technology*, 2nd ed.; John Wiley & Sons: New York, NY, USA, 1966; pp. 31–52.

21. Chen, F.; Zhao, J.; Hidaka, H. Highly selective deethylation of rhodamine B: Adsorption and photooxidation pathways of the dye on the TiO$_2$/SiO$_2$ composite photocatalyst. *Int. J. Photoenergy* **2003**, *5*, 209–217.

22. Colón, G.; Murcia López, S.; Hidalgo, M.C.; Navío, J.A. Sunlight highly photoactive Bi$_2$WO$_6$–TiO$_2$ heterostructures for rhodamine B degradation. *Chem. Commun.* **2010**, *46*, 4809–4811.

23. Chen, J.; Kou, S.; Poon, C. Photocatalytic cement-based materials: Comparison of nitrogen oxides and toluene removal potentials and evaluation of self-cleaning performance. *Build. Environ.* **2011**, *46*, 1827–1833.
24. Chen, J.; Kou, S.; Poon, C. Hydration and properties of nano-TiO_2 blended cement composites. *Cem. Concr. Compos.* **2012**, *34*, 642–649.

Transparent, Adherent, and Photocatalytic SiO$_2$-TiO$_2$ Coatings on Polycarbonate for Self-Cleaning Applications

Sanjay S. Latthe, Shanhu Liu, Chiaki Terashima, Kazuya Nakata and Akira Fujishima

Abstract: Photocatalytic TiO$_2$ coatings are famously known for their excellent self-cleaning behavior, where very thin water layer formed on the superhydrophilic surface can easily wash-off the dirt particles while flowing. Here we report the preparation of the optically transparent, adherent, highly wettable towards water and photocatalytic SiO$_2$-TiO$_2$ coatings on polycarbonate (PC) substrate for self-cleaning applications. The silica barrier layer was applied on UV-treated PC substrate before spin coating the SiO$_2$-TiO$_2$ coatings. The effect of different vol% of SiO$_2$ in TiO$_2$ and its influence on the surface morphology, mechanical stability, wettability, and photocatalytic properties of the coatings were studied in detail. The coatings prepared from 7 vol% of SiO$_2$ in TiO$_2$ showed smooth, crack-free surface morphology and low surface roughness compared to the coatings prepared from the higher vol% of SiO$_2$ in TiO$_2$. The water drops on this coating acquires a contact angle less than 10° after UV irradiation for 30 min. All the coatings prepared from different vol% (7 to 20) of SiO$_2$ in TiO$_2$ showed high transparency in the visible range.

Reprinted from *Coatings*. Cite as: Latthe, S.S.; Liu, S.; Terashima, C.; Nakata, K.; Fujishima, A. Transparent, Adherent, and Photocatalytic SiO$_2$-TiO$_2$ Coatings on Polycarbonate for Self-Cleaning Applications. *Coatings* **2014**, *4*, 497-507.

1. Introduction

TiO$_2$ is one of the most widely researched photocatalytic semiconductor materials to date. When irradiated with UV light, TiO$_2$ can decompose the organic pollutants present on its surface, and in addition the surface turns to be highly hydrophilic [1]. Water spread out instantaneously by forming a thin layer on the superhydrophilic TiO$_2$ thin films and steadily carries away dust particles from the surface while flowing (Figure 1). Photocatalytic superhydrophilic thin films with excellent self-cleaning properties are receiving much research attention worldwide because such smart surfaces can save time and reduce maintenance costs [2]. Today, the optically transparent, superhydrophilic and photocatalytic TiO$_2$ thin films are finding increasing industrial applications. Many reports on application of self-cleaning TiO$_2$ thin films on glass substrates are available in the literature [3–6]. An attractive soft polymer material, polycarbonate (PC) has great demand in the optical industry. PC can replace heavy glass in the optical and electronic industries in the coming future due to its extremely low weight, durability, low-cost, and optical transparency [7]. Many attempts have been made to improve the scratch resistance and UV degradation property of the PC by coating the surface by numerous photocatalytic materials with suitable binders.

Figure 1. Schematic showing the self-cleaning phenomena on superhydrophilic surface.

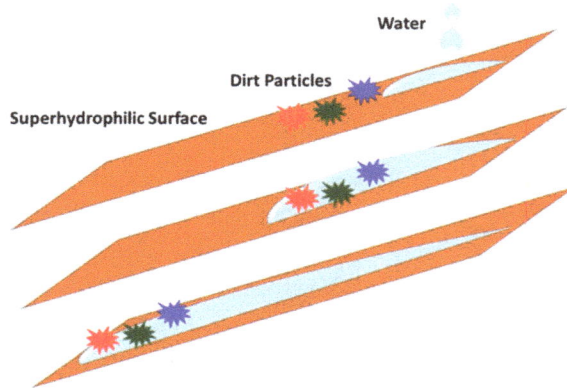

Hwang and coworkers [8] have prepared mechanically durable and optically transparent organic-inorganic SiO_2-TiO_2 nanocomposite coatings on PC by simple spray coating. The coating solution was prepared by incorporating nanosized TiO_2 sol into the silica polymeric sol. This nanocomposite coating also protected PC from photo-degradation under UV illumination. Lam *et al.* [9] reported the comparative study on the influence of NaOH etching and UVC irradiation on the mechanical stability of the TiO_2 thin films prepared on PC. They observed adherent TiO_2 film on UVC irradiated PC than NaOH treated PC due to increase in –OH and –COOH groups on the UVC-treated PC. However, the strong hydrophilic, antifogging, photocatalytic and self-cleaning properties were observed on the TiO_2 film coated on NaOH treated PC. Recently, Razan Fateh *et al.* have prepared optically transparent, highly hydrophilic, mechanically stable and photocatalytic TiO_2/SiO_2 [7,10], and TiO_2/ZnO coating [11] on PC for self-cleaning applications. In the present research work, we prepared transparent, adherent, highly wettable and self-cleaning SiO_2-TiO_2 coating on PC by simple spin coating. The coating showed faster wettability switching (hydrophilic to superhydrophilic) after 30 min. of UV irradiation. The effect of different vol% of SiO_2 sol in TiO_2 and its influence on the surface morphology, mechanical stability, wettability, and photocatalytic properties of the coatings were studied in detail.

2. Results and Discussion

2.1. Surface Microstructure, Roughness and Chemical Composition of the Coatings

The surface morphologies of the coatings prepared with different vol% of SiO_2 in TiO_2 are shown in Figure 2. The SiO_2-TiO_2 coating prepared with 7 vol% of SiO_2 in TiO_2 showed uniform, crack-free and smooth morphology (Figure 2a). However, significant cracks on the entire coating surface were observed with increase in SiO_2 concentration in TiO_2 (Figure 2b–d). Some small cracks were started to appear on the coating prepared from 10 vol% of SiO_2 in TiO_2 (Figure 2b), and these cracks goes bigger with increase in vol% of SiO_2. For the coating prepared with 20 vol% of SiO_2 in TiO_2, the cracks on the surface were abundant and coating material was popped out detaching from the substrate (Figure 2d). In the case of sol-gel coated films, during drying process, the capillary forces might have generated which provides cracks on the surface [12]. The surface roughness of the coatings were also studied (Figure 3). The SiO_2-TiO_2 coating prepared with 7 vol% of SiO_2 in TiO_2 showed the surface roughness of 67 nm (Figure 3a), whereas the surface roughness increases drastically to 98, 191 and 287 nm for the coating prepared from 10, 15 and 20 vol% of SiO_2 in TiO_2 (Figure 3b–d), respectively. This drastic increase in surface roughness may due to the increased density of cracks on the coating surface.

Figure 2. Field Emission Scanning Electron Microscope (FE-SEM) images of the coatings prepared from (**a**) 7; (**b**) 10; (**c**) 15; and (**d**) 20 vol% of SiO_2 in TiO_2.

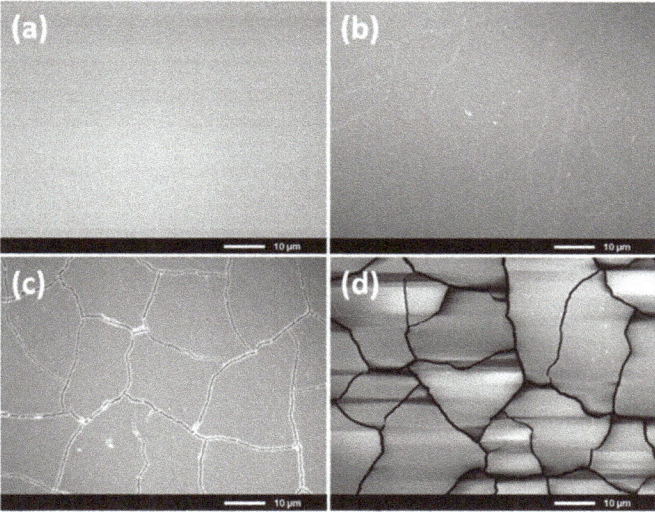

Figure 3. 3D Laser microscope images of the coatings prepared from (**a**) 7; (**b**) 10; (**c**) 15; and (**d**) 20 vol% of SiO_2 in TiO_2.

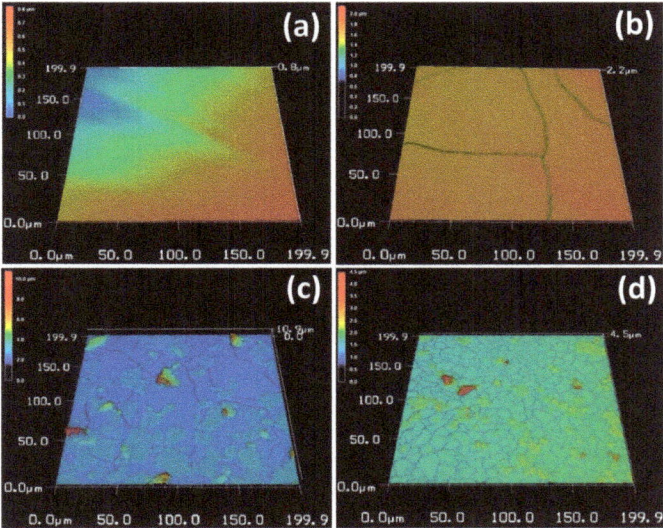

Figure 4 shows the FT-IR spectra of the coating prepared from 7 vol% of SiO_2 in TiO_2. The peaks observed in between 500 and 900 cm^{-1} can be attributed to the characteristic vibrational modes of TiO_2 [13]. A peak observed near 954 cm^{-1} is associated with Si-O-Ti vibration [14]. The absorption peak at 1118 cm^{-1} confirms Si-O-Si linkage [14]. The absorption peaks near 3342 and 1630 cm^{-1} can be attributed to the presence of stretching and bending vibrations of hydroxyl groups, respectively [15].

Figure 4. FT-IR spectra of the coatings prepared from 7 vol% of SiO_2 in TiO_2.

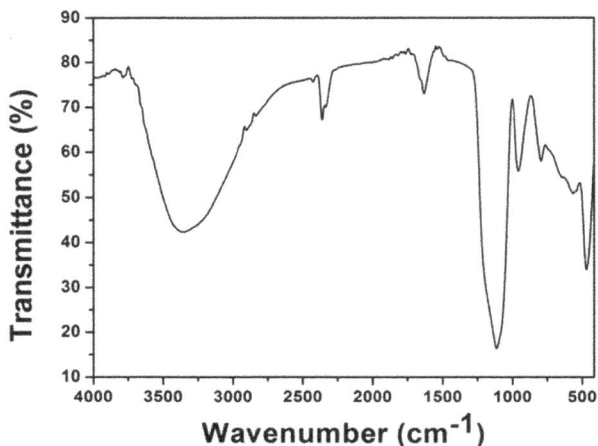

2.2. Superhydrophilic, Photocatalytic and Optical Properties of the Coatings

The wettability transition of the SiO_2-TiO_2 coating after UV illumination was studied. The prepared SiO_2-TiO_2 coatings were illuminated by UV light (365 nm, 2 mW/cm^2). The UV illumination creates structural changes in TiO_2 (transformation of Ti^{4+} sites to Ti^{3+} sites) [16] and oxidizes the organic contaminants present on the surface of TiO_2, which effectively transforms the wettability of the TiO_2 surface towards more hydrophilic [17]. This is called as photo-induced hydrophilicity on TiO_2 surface [18]. The water contact angles (WCA) on the coatings before and after exposure to the UV light were measured. The water drop volume of 5 µL was used to measure the water contact angles on the coating surface. The effect of UV exposure time on the wettability of the coatings was also studied. Figure 5 shows the wettability of the coatings prepared from 7 and 20 vol% of SiO_2 in TiO_2, before and after UV irradiation. Before UV illumination, the coating prepared from 7 and 20 vol% of SiO_2 in TiO_2 showed the WCA of 23° and 28°, respectively. After UV illumination of 30 min, the water drop immediately spread on the surface and the WCA drastically decreased to 8° on the coating prepared from 7 vol% of SiO_2 in TiO_2 and even for longer UV illumination time (5 h), the WCA remained in the range of 7°. The coating prepared from 20 vol% of SiO_2 in TiO_2 also showed decrease in WCA in the range of 10°, after longer UV illumination time. This slightly higher WCA in case of the coating prepared from 20 vol% of SiO_2 in TiO_2 is due to relatively high surface roughness provided by significant density of cracks present on the surface. Even after placing the UV irradiated coatings in dark for 3 months, the coatings showed stable wetting properties with WCA measured well below 10°. Figure 6 shows the optical photograph of water drops on bare PC substrate and the coating prepared from 7 vol% of SiO$_2$ in TiO$_2$ after 30 min. UV irradiation. Some of the water drops were colored blue using Methylene Blue for better visualization of water drops. The water drops on bare PC substrate maintain the contact angle of ~84°, whereas the water drops spreads on UV irradiated SiO$_2$-TiO$_2$ coating, confirming high affinity towards the water. In the case of SiO$_2$-TiO$_2$ coatings prepared by Fateh et al. [7], it needed more than 700 h of UV (A) irradiation to switch the wettability of the coatings in the superhydrophilic range.

Figure 5. Wetting properties of the coatings prepared from 7 and 20 vol% of SiO$_2$ in TiO$_2$ (Insets shows the shape of water drops on coatings).

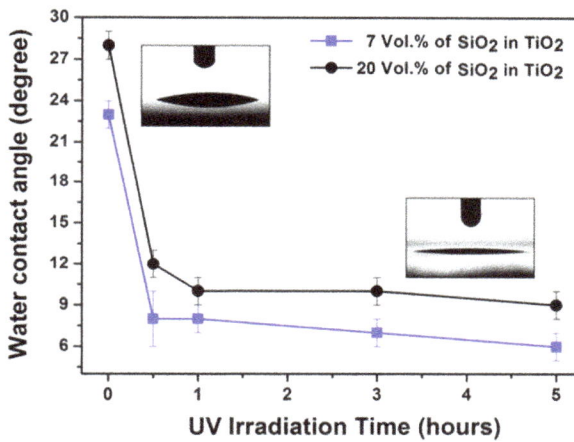

Figure 6. Optical photograph of water drops on bare PC substrate (**a**) and coating prepared from 7 vol% of SiO$_2$ in TiO$_2$ after 30 min. UV irradiation (**b**).

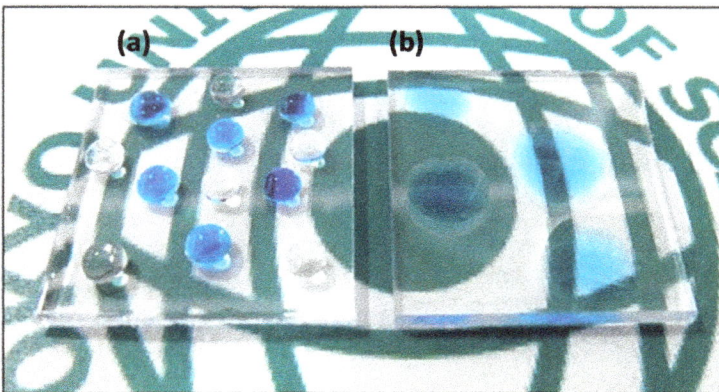

The TiO$_2$ is famously known for its excellent photocatalytic property, as it can degrade organic contamination under the illumination of UV light. We studied the photocatalytic degradation of the ODS monolayers after UV light illumination (365 nm, 2 mW/cm^2) by means of water contact angle measurements. At first, the bare PC substrate and coating prepared from 7 vol% of SiO$_2$ in TiO$_2$

were irradiated with UV light (365 nm, 2 mW/cm^2) for 1 h to make them superhydrophilic and hydrophobic ODS self-assembled monolayers (SAMs) were applied on them through vapor phase. To employ ODS SAMs on the surface, 2 mL of ODS in beaker was kept in the closed metal box containing the samples at 100 °C for 24 h. After application of ODS SAMs on the surface, the WCA on bare PC substrate and coating prepared from 7 vol% of SiO$_2$ in TiO$_2$ showed 95° and 50°, respectively (Figure 7). The ODS treated bare PC showed almost no change in WCA after 120 min of UV illumination, whereas the coating prepared from 7 vol% of SiO$_2$ in TiO$_2$ showed significant decrease in WCA as a function of UV illumination time and the surface becomes superhydrophilic after 120 min of UV illumination (Figure 7) confirming photocatalytic degradation of the ODS monolayers.

Figure 7. Wettability change on octadecyltrichlorosilane (ODS)-treated SiO$_2$-TiO$_2$ coatings under UV illumination.

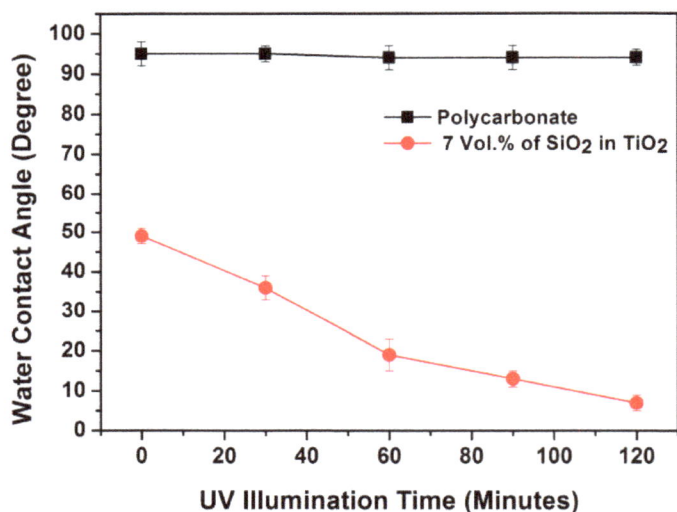

Optical transparency is prerequisite for the application of self-cleaning coating on transparent glass or plastic surface. Figure 8 shows the optical transmission of SiO$_2$-TiO$_2$ coatings prepared on the PC substrate from different vol% of SiO$_2$ in TiO$_2$. All the coatings are highly transparent and showed the optical transmittance values above 85% over the entire visible wavelength range. The optical transmission was gradually decreased from 89% to 85% with an increase in SiO$_2$ concentration in TiO$_2$. The increased surface roughness is responsible for the slight loss in optical transmission in the visible range. The SiO$_2$-TiO$_2$ coatings prepared by Hwang et al. [8] showed an optical transmission of ~87%, whereas all the coatings prepared by Fateh et al. [7,10] showed an optical transmission of >92%.

Figure 8. Optical transmission spectra of the SiO_2-TiO_2 coatings.

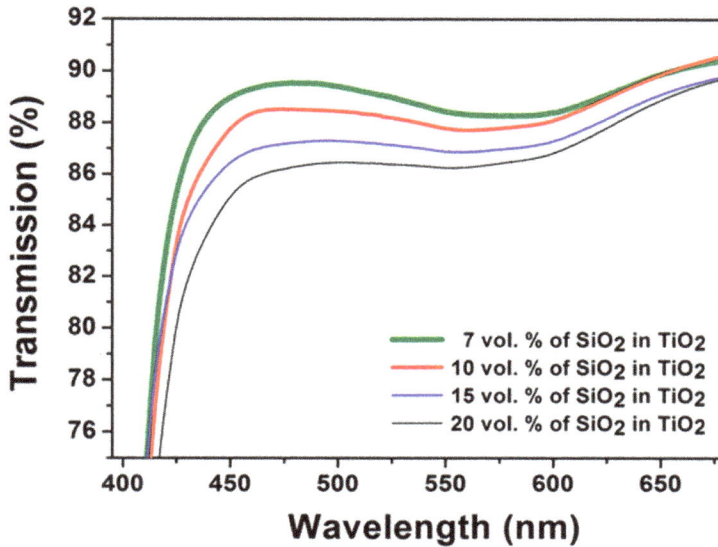

2.3. Mechanical Properties of the Coatings

The mechanical durability of the coatings is a very important criterion for industrial applications. The mechanical stability of the prepared SiO_2-TiO_2 coatings is depicted in Figure 9. The bare PC substrates get easily scratched at the applied force of less than 4.2 mN. For the coating prepared from 7 vol% of SiO_2 in TiO_2, the coating material was removed at the applied force of ~16.1 mN (Figure 9a), whereas this force was decreased to 12.2, 10.9 and 9.8 mN for the coating prepared from 10, 15 and 20 vol% of SiO_2 in TiO_2 (Figure 9b–d), respectively. The smooth, crack-free morphology and lower surface roughness on the coating resulted in the stable and adhesive coating on the PC substrate. Even rubbed by the fingers, the coating material was not easily removed. This is ascribed to the formation of an intermediate layer connecting the TiO_2 nanoparticles and the silica network. However, in the case of the coating prepared from higher concentration of SiO_2 in TiO_2, due to significant cracks and increased surface roughness, the coating material could be removed at relatively lower applied force.

Figure 9. Adhesion test performed on the coatings prepared from (**a**) 7; (**b**) 10; (**c**) 15; and (**d**) 20 vol% of SiO_2 in TiO_2.

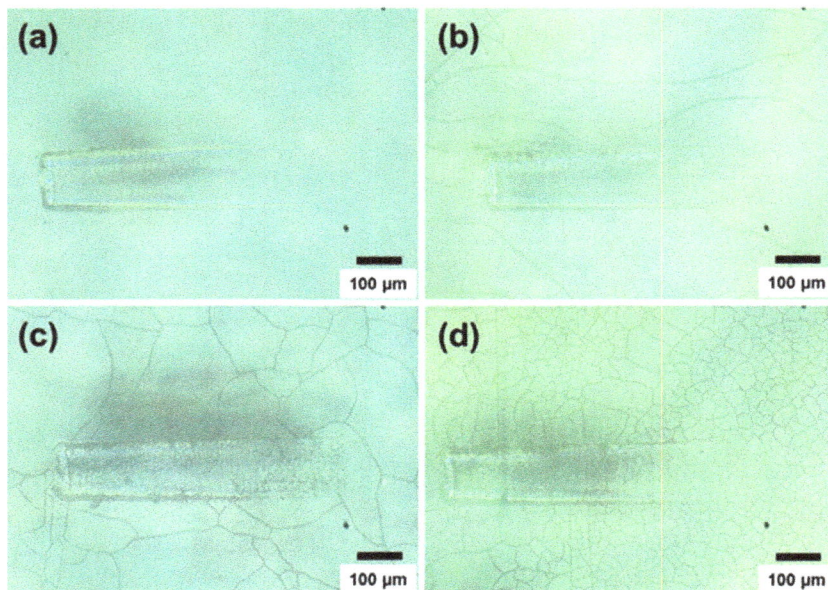

3. Experimental Section

3.1. Materials

Tetraethylorthosilicate (TEOS) and Octadecyltrichlorosilane (ODS) were purchased from Sigma Aldrich (St. Louis, MO, USA). Ethanol (99.5%) and nitric acid (69%) were purchased from Wako Pure Chemical Industries Ltd. (Kanto region, Japan). PC substrates (ECK, 100UU) were bought from Sumitomo Bakelite Co. Ltd. (Akita, Japan). The commercially available TiO_2 nanoparticle solution (NRC-300C) was purchased from Nippon Soda Co. Ltd. (Tokyo, Japan).

3.2. Self-Cleaning Coating on PC Substrates

PC substrates are known to be hydrophobic in nature and so the adhesion is not strong between the coating material and the PC substrate [19]. The hydrophobic nature of PC substrates can be transformed to strongly hydrophilic by UV irradiation due to occurrence of photo-Fries reaction on the surface [20]. At first, PC substrates were gently cleaned by using detergent and water and kept for ultrasonication in double distilled water for 30 min. After drying at room temperature, the PC substrates were illuminated by UV light (365 nm, 2 mW/cm^2) for 2 h. The PC substrates can photocatalytically degrade, if the TiO_2 coating is applied directly on the PC substrates. For this reason, the sol-gel processed silica layer was applied on the PC substrates prior to TiO_2 coating. A silica sol was prepared by adding 1.5 mL TEOS in 4 mL ethanol and this mixture was hydrolysed by adding 3 mL of double distilled water with 1 mL of nitric acid. This alcosol was stirred overnight and spin-coated on UV-treated PC substrates at 2000 rpm. This silica coating was annealed at 100 °C for 2 h. The above prepared silica sol was mixed at different vol% (7, 10, 15 and 20 vol%) in commercially available TiO$_2$ nanoparticle solution and spin-coated on SiO$_2$ pre-coated PC substrates with 2000 rpm and annealed in oven at 110 °C for 6 h. No change in PC substrate was observed after annealing at 110 °C, however for annealing above 120 °C, PC substrate start to bend in irregular shape due to softening. The schematic of coating preparation steps are shown in Figure 10.

Figure 10. Schematic showing coating preparation steps.

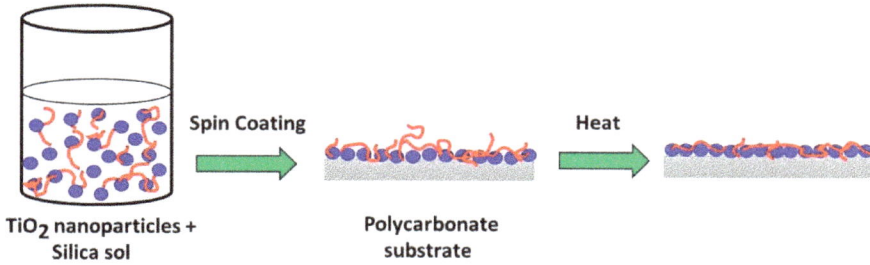

TiO$_2$ nanoparticles + Silica sol

Spin Coating

Polycarbonate substrate

Heat

3.3. Characterization Techniques

The surface morphology was studied by Field Emission Scanning Electron Microscope (FE-SEM, JEOL, JSM-7600F; Tokyo, Japan). The surface roughness of the prepared coatings was evaluated by laser microscopy (KEYENCE, VK-X200 series; Itasca, IL, USA). The chemical composition was studied by Fourier transform infrared spectroscopy (FT-IR, JASCO, FT/IR-6100; Tokyo, Japan). The water contact angles (WCA) were measured at six different locations on same sample by using contact angle meter (KYOWA, Drop Master; Saitama, Japan) and average value is reported as a final contact angle value. The optical transmission of the coatings was measured by UV-VIS spectrophotometer (JASCO, V-670; Tokyo, Japan). The coatings were illuminated by UV light (365 nm, 2 mW/cm^2). The UV lamps were purchased from TOSHIBA (FL10BLB; Tokyo, Japan) and assembled with proper electric supply inside the wooden box covered by thick black cloth. The adhesion of the coating material on PC substrate was checked at five different spots by using Scratch tester (Nano-Layer Scratch Tester, CSR-2000; Rhesca, Tokyo, Japan). The adhesion of the coatings on PC substrate and the maximum force required to remove the coating material was calculated. The cantilever was moved from right to left side in contact with the coating surface. The force was gradually increased while moving towards left side. The maximum force at which the coating material was removed from the PC substrate was noted as critical force to damage the coating. The optical photographs were recorded using Canon Digital Camera (G 15 series).

4. Conclusions

The mechanically durable, optically transparent, photocatalytically active and superhydrophilic SiO$_2$-TiO$_2$ coatings are successfully prepared on PC substrates for self-cleaning applications. The uniform, crack-free coatings prepared from 7 vol% of SiO$_2$ in TiO$_2$ showed higher optical transmission in the visible range, strong superhydrophilicity and good scratch resistance properties.

Such coatings can be applied on light-weight window and door polycarbonates for excellent self-cleaning applications.

Acknowledgments

Authors Sanjay S. Latthe (P13067) and Shanhu Liu (P12345) are grateful for the financial support provided by the Japan Society for the Promotion of Science (JSPS), Japan, under Postdoctoral Fellowship for Foreign Researchers.

Author Contributions

Sanjay S. Latthe conceived the idea and designed the structure of article. Sanjay S. Latthe and Shanhu Liu carried out the experiments and wrote the original manuscript and Chiaki Terashima, Kazuya Nakata and Akira Fujishima participated in the discussions and helped to revise it.

Conflicts of Interest

The authors declare no conflict of interest.

References

1. Sakai, N.; Fukuda, K.; Shibata, T.; Ebina, Y.; Takada, K.; Sasaki, T. Photoinduced hydrophilic conversion properties of titania nanosheets. *J. Phys. Chem. B* **2006**, *110*, 6198–6203.
2. Blossey, R. Self-cleaning surfaces—virtual realities. *Nat. Mater.* **2003**, *2*, 301–306.
3. Weng, K.-W.; Huang, Y.-P. Preparation of TiO_2 thin films on glass surfaces with self-cleaning characteristics for solar concentrators. *Surf. Coat. Technol.* **2013**, *231*, 201–204.
4. Xi, B.; Verma, L.K.; Li, J.; Bhatia, C.S.; Danner, A.J.; Yang, H.; Zeng, H.C. TiO_2 thin films prepared via adsorptive self-assembly for self-cleaning applications. *ACS Appl. Mater. Interfaces* **2012**, *4*, 1093–1102.
5. Euvananont, C.; Junin, C.; Inpor, K.; Limthongkul, P.; Thanachayanont, C. TiO_2 optical coating layers for self-cleaning applications. *Ceram. Int.* **2008**, *34*, 1067–1071.
6. Lai, Y.; Tang, Y.; Gong, J.; Gong, D.; Chi, L.; Lin, C.; Chen, Z. Transparent superhydrophobic/superhydrophilic TiO_2-based coatings for self-cleaning and anti-fogging. *J. Mater. Chem.* **2012**, *22*, 7420–7426.
7. Fateh, R.; Dillert, R.; Bahnemann, D. Preparation and characterization of transparent hydrophilic photocatalytic TiO_2/SiO_2 thin films on polycarbonate. *Langmuir* **2013**, *29*, 3730–3739.
8. Hwang, D.K.; Moon, J.H.; Shul, Y.G.; Jung, K.T.; Kim, D.H.; Lee, D.W. Scratch resistant and transparent UV-protective coating on polycarbonate. *J. Sol Gel Sci. Technol.* **2003**, *26*, 783–787.
9. Lam, S.; Soetanto, A.; Amal, R. Self-cleaning performance of polycarbonate surfaces coated with titania nanoparticles. *J. Nanopart. Res.* **2009**, *11*, 1971–1979.
10. Fateh, R.; Ismail, A.A.; Dillert, R.; Bahnemann, D.W. Highly active crystalline mesoporous TiO_2 films coated onto polycarbonate substrates for self-cleaning applications. *J. Phys. Chem. C* **2011**, *115*, 10405–10411.

11. Fateh, R.; Dillert, R.; Bahnemann, D. Self-cleaning properties, mechanical stability, and adhesion strength of transparent photocatalytic TiO_2-ZnO coatings on polycarbonate. *ACS Appl. Mater. Interfaces* **2014**, *6*, 2270–2278.

12. Hatton, B.; Mishchenko, L.; Davis, S.; Sandhage, K.H.; Aizenberg, J. Assembly of large-area, highly ordered, crack-free inverse opal films. *Proc. Natl. Acad. Sci. USA* **2010**, *107*, 10354–10359.

13. Khanna, P.K.; Singh, N.; Charan, S. Synthesis of nano-particles of anatase-TiO_2 and preparation of its optically transparent film in PVA. *Mater. Lett.* **2007**, *61*, 4725–4730.

14. Kumar, D.A.; Shyla, J.M.; Xavier, F.P. Synthesis and characterization of TiO_2/SiO_2 nano composites for solar cell applications. *Appl. Nanosci.* **2012**, *2*, 429–436.

15. Rao, A.V.; Latthe, S.S.; Nadargi, D.Y.; Hirashima, H.; Ganesan, V. Preparation of MTMS based transparent superhydrophobic silica films by sol–gel method. *J. Colloid Interface Sci.* **2009**, *332*, 484–490.

16. Shultz, A.N.; Jang, W.; Hetherington, W.M., III; Baer, D.R.; Wang, L.-Q.; Engelhard, M.H. Comparative second harmonic generation and X-ray photoelectron spectroscopy studies of the UV creation and O_2 healing of Ti^{3+} defects on (110) rutile TiO_2 surfaces. *Surf. Sci.* **1995**, *339*, 114–124.

17. Wang, R.; Hashimoto, K.; Fujishima, A.; Chikuni, M.; Kojima, E.; Kitamura, A.; Shimohigoshi, M.; Watanabe, T. Light-induced amphiphilic surfaces. *Nature* **1997**, *388*, 431–432.

18. Anandan, S.; Rao, T. N.; Sathish, M.; Rangappa, D.; Honma, I.; Miyauchi, M. Superhydrophilic graphene-loaded TiO_2 thin film for self-cleaning applications. *ACS Appl. Mater. Interfaces* **2012**, *5*, 207–212.

19. Aslan, K.; Holley, P.; Geddes, C.D. Metal-enhanced fluorescence from silver nanoparticle-deposited polycarbonate substrates. *J. Mater. Chem.* **2006**, *16*, 2846–2852.

20. Rivaton, A. Recent advances in bisphenol—A polycarbonate photodegradation. *Polym. Degrad. Stab.* **1995**, *49*, 163–179.

Self-Cleaning Photocatalytic Polyurethane Coatings Containing Modified C$_{60}$ Fullerene Additives

Jeffrey G. Lundin, Spencer L. Giles, Robert F. Cozzens and James H. Wynne

Abstract: Surfaces are often coated with paint for improved aesthetics and protection; however, additional functionalities that impart continuous self-decontaminating and self-cleaning properties would be extremely advantageous. In this report, photochemical additives based on C$_{60}$ fullerene were incorporated into polyurethane coatings to investigate their coating compatibility and ability to impart chemical decontaminating capability to the coating surface. C$_{60}$ exhibits unique photophysical properties, including the capability to generate singlet oxygen upon exposure to visible light; however, C$_{60}$ fullerene exhibits poor solubility in solvents commonly employed in coating applications. A modified C$_{60}$ containing a hydrophilic moiety was synthesized to improve polyurethane compatibility and facilitate segregation to the polymer–air interface. Bulk properties of the polyurethane films were analyzed to investigate additive–coating compatibility. Coatings containing photoactive additives were subjected to self-decontamination challenges against representative chemical contaminants and the effects of additive loading concentration, light exposure, and time on chemical decontamination are reported. Covalent attachment of an ethylene glycol tail to C$_{60}$ improved its solubility and dispersion in a hydrophobic polyurethane matrix. Decomposition products resulting from oxidation were observed in addition to a direct correlation between additive loading concentration and decomposition of surface-residing contaminants. The degradation pathways deduced from contaminant challenge byproduct analyses are detailed.

Reprinted from *Coatings*. Cite as: Lundin, J.G.; Giles, S.L.; Cozzens, R.F.; Wynne, J.H. Self-Cleaning Photocatalytic Polyurethane Coatings Containing Modified C$_{60}$ Fullerene Additives. *Coatings* **2014**, *4*, 614-629.

1. Introduction

Chemical toxicants such as pesticides and toxic industrial chemicals have the potential to contaminate material surfaces for extended periods of time. Often the natural attenuation of toxic substances through evaporation and degradation in ambient environments occurs slowly due to low vapor pressure, poor solubility, and resistance to hydrolysis [1]. Due to this persistence, surfaces on which toxic chemicals reside pose human exposure risks that require application of decontamination solutions to completely render a surface safe. However, decontamination solutions and procedures are cumbersome, expensive, and often damaging to the contaminated substrate [2].

Polymeric coatings are typically applied to many surfaces to improve aesthetics and provide corrosion or weathering protection; therefore, they provide an ideal substrate to incorporate coating additives to impart continuous self-decontaminating behavior at the surface and reduce subsequent contamination. A minimal loading concentration of additive is ideal so that the properties beneficial for the originally intended purpose of the polymer coating are maintained.

Specifically, polyurethanes are of the broadest interest owing to their properties such as chemical resistance and durability [3].

Several recent research developments have investigated the incorporation of novel reactive additives into various urethane coating formulations in attempts to create coatings that self-decontaminate. Antimicrobial coatings have been successfully created by imparting additives such as nonionic biocides [4], quaternary ammonium biocides [5,6], surface concentrating biocides [7], functionalized coatings [8,9], and antimicrobial peptides [10]. While these are successful biocidal additives, less success has been achieved in chemical decontaminating coatings. One reason is that additives for chemical decontamination are limited to only a few modes of action such as absorption, hydrolysis, and oxidation. Of these, oxidation offers the greatest potential to completely detoxify a broad spectrum of chemical contaminants [11].

C_{60} fullerene molecules have also been observed to exhibit intriguing photochemical properties, including oxidative capabilities, which hold exciting potential for development of a self-decontaminating coating [12–15]. Upon exposure to visible light, C_{60} fullerene is first excited to its singlet state C_{60} ($^1C_{60}$), which then through intersystem crossing (ISC), forms the triplet state species of C_{60} ($^3C_{60}$). $^3C_{60}$ has a lifetime on the order of µs whereas $^1C_{60}$ exhibits a lifetime of several ns [12,16–18]. This triplet state species of fullerene has the ability to convert ground state triplet oxygen ($^3\Sigma_g^-$) into singlet oxygen ($^1\Delta_g$), a reactive oxygen species (ROS) [13,19]. The combination of a high quantum yield [13] and low rate of degradation of C_{60} fullerene by ROS [12] make this molecule extremely attractive as a photo-active coating additive.

Extensive studies have been conducted to analyze and characterize the photosensitivity of C_{60} in solution with varying degrees of success [16,19–24]. Various photosensitizers have been shown to exhibit antimicrobial activity when incorporated into polyurethane coating systems under specific conditions [25–27]. Similarly, antiviral systems have successfully been developed with the incorporation of fullerene as a solid-phase photosensitizer into biological fluid [28]. Recently, fullerenes modified with intercage constituents have displayed a remarkable ability to produce singlet oxygen as well as antimicrobial activity in polymeric adhesive films [29]. However, insertion of intercage constituents into fullerenes introduces additional cost and complexity that may be avoided by utilizing neat C_{60}.

Incorporation of C_{60} fullerene into polymer matrices has been investigated for applications ranging from photovoltaics [30] to augmentation of polymer mechanical properties [31]. Covalent incorporation of C_{60} fullerene into polymers offers controlled distribution and reduced leaching, albeit often at the sacrifice of photoactivity [32,33]. In contrast, non-covalent incorporation of C_{60} fullerene offers simplified formulation and unaffected photophysical properties [34]. Furthermore, non-covalent incorporation of an amphiphilic fullerene species affords the potential for surface segregating photoactive additives. While synthetic modification of C_{60} into an amphiphilic species most likely will affect the photophysical properties of C_{60}, increased concentration of a photoactive additive at the surface of a polymer due to its amphiphilic character should improve the decomposition of surface-residing chemical contaminants.

It can be assumed that when incorporated into a polymer matrix, the photoactivity of fullerene may be reduced due to a lack of molecular oxygen available to the fullerene molecule if it is encapsulated into the bulk of the polymeric coating. However, if one is able to overcome this limitation, significant activity should remain at the coating–air interface. It should then be expected that the production of ROS would result and subsequently react with any contamination that may be on the surface. Furthermore, an amphiphilic additive that automatically segregates to the polymer–air interface during a film cure would improve decontaminating efficiency. The hypothesis proposed herein is that the fullerene contained in the coating produces singlet oxygen from the atmosphere by the aforementioned mechanism and subsequently reacts with undesired contamination analytes that are present on the surface. If such analytes are hazardous, such as the case of pesticides or chemical warfare agents, then the action of the additive in the coating should reduce the hazard and subsequently present a surface free from contamination.

2. Experimental Section

2.1. Materials

All purchased chemicals were reagent grade and were used without further purification. The purchase of 2-chloroethyl phenyl sulfide was made at Sigma–Aldrich (St. Louis, MO, USA). Demeton-S was purchased from Chem Service (West Chester, PA, USA). Refined C_{60} fullerene was purchased from MER Corporation (Tucson, AZ, USA).

2.2. Synthesis of EO₃–C₆₀

Covalent attachment of triethylene glycol monomethyl ether to C_{60} fullerene through azide addition was performed following a previously described method [35]. Product was confirmed by ^{1}H NMR performed in CDCl₃ on a Bruker 300 MHz nuclear magnetic resonance spectrometer with a TMS internal standard. Characterization data is consistent with that previously reported [35].

2.3. Polymer Preparation

A commercial polyurethane resin, Tecoflex EG-100A (Lubrizol Advanced Materials, Inc., Gastonia, NC, USA), was employed to investigate behavior of additives in controlled polymer solutions. Polymer solutions were prepared (0.1 g polymer in 1 mL chloroform), to which photoactive additives, C_{60} and EO_3–C_{60}, were incorporated. After addition of additives, the solutions were vigorously vortexed to ensure complete mixing. After allowing for air bubbles to escape, polymer solutions were poured out into aluminum pans, loosely covered, and allowed to dry. Films with final additive concentrations of 0.25, 0.5, 1.0, 2.5, and 5.0 wt% were prepared.

2.4. Coating Characterization

Free coatings were peeled from aluminum backing and weighed prior to analysis on a TA Instruments (New Castle, DE, USA) DSC Q20 differential scanning calorimeter. Under a nitrogen flow of 50 mL/min, the DSC was first equilibrated to −70 °C. The temperature was then ramped from −70 °C to 170 °C at a rate of 20 °C/min. The procedure was repeated for a second scan with a ramp rate of 10 °C/min. Glass transition temperature (T_g) measurements were calculated based on the second scan using Universal Analysis 2000 software. The second scan was used for this value to eliminate any contamination of entrapped volatile and low molecular weight byproducts, as well as demonstrate hysteresis. Thermogravimetric analysis (TGA) was performed on a TA Instruments Q50 TGA employing heating rates of 10 °C/min under a N_2 atmosphere. A Thermo Scientific Nicolet 6700 FTIR (Thermo Scientific, Waltham, MA, USA) equipped with a diamond crystal ATR attachment was utilized for film analysis. Diffuse reflectance was utilized in the characterization of neat C_{60} and EO_3–C_{60}. X-ray diffraction measurements were performed using a Rigaku SmartLab X-ray Diffractometer (XRD, Rigaku, Tokyo, Japan). The SmartLab XRD was equipped with a Cu anode operating at 3 kW generating Cu Kα radiation. Measurements were taken with Bragg–Brentano Optics and a D/Tex Detector for 2-Theta measurements from 15° to 40°.

Contact angle measurements were performed using a VCA Optima by AST Products, Inc. (Billerica, MA, USA) employing the sessile drop technique. Triple-distilled water was employed as a probe liquid, of which at least three replicate measurements were made for each sample. 3D laser confocal microscopy was performed on an Olympus LEXT 3D measuring laser microscope OLS4000. Surface roughness measurements were performed employing an 80 µm cutoff wavelength (λ_c).

2.5. Contaminant Challenge

The prepared films were subjected to surface decontamination challenges against the chemical analytes presented in Figure 1, Demeton-S (**1**) and 2-chloroethyl phenyl sulfide (CEPS) (**2**), across a range of simulated environmental conditions. In general, a 2.0 µL micropipette was used to apply 1.0 µL of analyte to each sample (2 cm²) placed in a transparent Eppendorf tube. Each Eppendorf tube was then sealed and allowed to incubate for determined period of time in controlled conditions (darkness or simulated daylight). For photochemical reactions, a custom-built temperature controlled photochemical reactor equipped with five F8T5D fluorescent bulbs emitting broad spectrum visible light (Figure 2) at 10,000 lux intensity was employed, which simulates overcast daylight exposure. All contaminant challenges, including photochemical challenges, were performed at 20 °C. After which, residual analyte and degradation byproducts were extracted from polymer films with 1 mL of acetonitrile solution containing 12.1 mM tetrahydronaphthalene as an internal standard. Samples were then placed into a 1.5 mL GC auto-sample vial with PTFE septa top and analyzed immediately. In addition to simulant work being performed in a fume hood, personal protective equipment consisting of nitrile gloves, lab coat, chemical safety goggles were employed at all times during handling of chemical simulants.

Figure 1. (1) Demeton-*S*; **(2)** *S*-vinyl degradation product; **(3)** CEPS; and **(4)** vinyl phenyl sulfoxide.

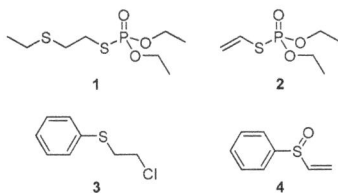

Figure 2. Emission spectrum of custom-built photochemical reactor.

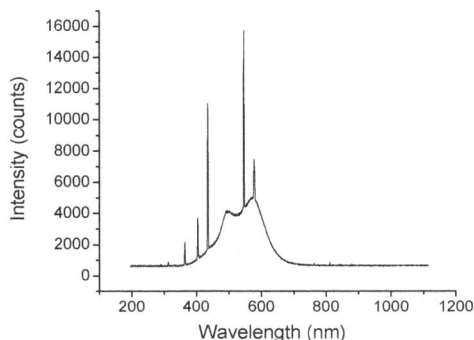

Gas chromatography/mass spectroscopy (GC/MS) was employed to quantify analyte degradation. The GC/MS system consisted of an Agilent 7890A gas chromatograph equipped with an Agilent 5975C mass selective detector operating in electron ionization mode and an Agilent 7693A autoinjector. The column utilized was an Agilent HP-5MS (5% phenyl) methylpolysiloxane film. The carrier gas was helium with a flow rate of 1 mL·min^{-1}. The injection volume was 1 μL with a split injection ratio of 20:1. The temperature program has an initial temperature of 100 °C for one minute, then 25 °C per min ramp to 130 °C followed by a 15 °C per min ramp to 250 °C with a one minute post run hold at 300 °C. The injection temperature, MS quad temperature, and source temperature were 300, 150 and 230 °C, respectively. The solvent delay was set at 1.5 min and detector was set to scan a mass range of 20 to 350 m/z. Prior to comparison, GC/MS results were normalized by dividing the analyte peak area by the peak area of the internal standard, tetrahydronaphthalene.

3. Results and Discussion

3.1. Additive–Polymer Compatibility

Additive–polymer compatibility and the effects of photoactive additive incorporation on Tecoflex film properties were investigated with a variety of methods. ATR-IR analysis was performed to investigate the effect of C_{60} incorporation on infrared signature and probe the chemical composition of the polymer film. ATR-IR analysis confirmed that the low loading concentrations employed in this study allowed the chemical signature of polymer film to be maintained despite increasing concentrations of C_{60} and EO_3–C_{60}.

Effects of additive incorporation on Tecoflex films on thermal stability were investigated with TGA. The incorporation of C_{60} and EO_3–C_{60} into Tecoflex films each resulted in an increased thermal stability (Table 1). The incorporation of EO_3–C_{60} in Tecoflex resulted in increased thermal stability (temperature at 10% loss) that scaled linearly with loading concentration. The increase in initial thermal stability suggests that stabilizing intermolecular interactions are occurring between EO_3–C_{60} and Tecoflex polymer. Generally, the incorporation of C_{60} into Tecoflex demonstrated a moderate increase in thermal stability. A weak inverse relationship was observed with increased loading of C_{60} and thermal stability, as opposed to the direct relationship of EO_3–C_{60} loading. This suggests that with increased loading concentration, additional C_{60} aggregates to other C_{60} instead of interacting with polymer matrix. Additionally, the lessened effect on thermal stability upon C_{60} incorporation compared to EO_3–C_{60} indicates that EO_3–C_{60} interacts more strongly than C_{60} with Tecoflex polymer matrix. The greater favorable interactions between EO_3–C_{60} and Tecoflex results from compatibility between the ethylene oxide tail of EO_3–C_{60} and the ether regions of the butane diol constituents of the Tecoflex monomer.

Table 1. Thermal properties of Tecoflex films.

Tecoflex Film	10% Loss (°C)	Mass Remaining (%) *
Control	300.55	0.2727
0.5% C_{60}	314.99	1.463
1.0% C_{60}	311.07	2.561
2.5% C_{60}	307.38	6.574
5.0% C_{60}	309.86	11.25
0.25% EO_3–C_{60}	312.61	0.6591
0.5% EO_3–C_{60}	319.44	0.7918
1.0% EO_3–C_{60}	325.63	1.513

* At 600 °C.

TGA analysis also indicated that incorporation of C_{60} into Tecoflex resulted in greater ultimate mass remaining than those loaded with EO_3–C_{60}. The alkoxy moiety of the EO_3–C_{60} is susceptible to thermal degradation at a greater temperature than the fullerene cage of C_{60}. Therefore, each comparable loading concentration of C_{60} (720 g/mol) and EO_3–C_{60} (881 g/mol) contains a 1.22-fold molar excess of C_{60} to EO_3–C_{60}. Thus, the thermal degradation of the alkoxy tail of EO_3–C_{60} below 600 °C results in only 82% remaining of the total loaded EO_3–C_{60}. Considering this, at 600 °C, there should be approximately a 1.5× excess remaining mass % of C_{60} relative to EO_3–C_{60} at comparable loading concentrations. Indeed, such an excess was observed upon comparison of the 1 wt% loadings.

Interestingly, the ultimate wt% for each Tecoflex film containing additives was greater than the additive wt% loading concentration (Table 1). The amount of mass remaining from the thermal degradation of Tecoflex films containing C_{60} at 600 °C corresponded to approximately double the C_{60} loading concentration. This indicates one of two possibilities: a quantity of Tecoflex polymer is strongly adhered to the surface of C_{60} fullerene through intermolecular forces that require greater than 600 °C to dissociate or aggregation of C_{60} results in thermally stabilized Tecoflex polymer trapped within the aggregate. However, the lower thermal degradation onset temperature of Tecoflex films containing C_{60} rather than EO_3–C_{60} revealed that intermolecular interactions were stronger for EO_3–C_{60}. The difference between the final mass remaining and loading concentration decreased with increased additive loading concentration of both C_{60} and EO_3–C_{60}, suggesting that with increased loading concentration, additional intermolecular C_{60} aggregation occurs instead of C_{60}–polymer matrix interactions. The deviation was greater for C_{60} than it was for EO_3–C_{60}, indicating that aggregation is more prominent in C_{60} than EO_3–C_{60}.

DSC was performed on all Tecoflex films to investigate effects of additive incorporation on glass transition temperature (T_g). The way in which additive incorporation affects T_g can afford insight into intermolecular interactions between the additive and polymer. Increase in T_g resulting from additive incorporation indicates increased intermolecular interactions that limit polymer mobility [36]. Unmodified Tecoflex exhibits a glass transition that spans a broad temperature range. Incorporation of additives, both C_{60} and EO_3–C_{60}, result in minor and insignificant effects on T_g, thus indicating that the integrity of the coating was preserved.

For both neat C_{60} and Tecoflex films containing higher concentrations of C_{60}, an endothermic transition was observed in the DSC thermograms at approximately -14 °C. The magnitude of the endothermic peak increases corresponding to C_{60} concentration (Figure 3). This endotherm corresponds to the phase transition of C_{60} crystals from simple cubic orientation ordering to face-centered cubic structure upon heating through the transition temperature [37,38]. The presence of this transition from Tecoflex films indicates a crystalline phase of C_{60} fullerene. C_{60} can only be in a crystalline phase when multiple fullerene molecules are in contact with one another, or aggregated, in a regular, repeated order. A polymeric film in which C_{60} fullerene was completely dispersed would not exhibit such crystalline phase transition. Furthermore, endotherm corresponding to the simple cubic to face-centered cubic phase transition was absent in the DSC analysis of neat EO_3–C_{60} and the Tecoflex films containing EO_3–C_{60}. Therefore, aggregation, or at least the formation of crystallites, of C_{60} is inhibited by the covalent modification of C_{60} with ethylene oxide tails. Furthermore, the amphiphilic character of the EO_3–C_{60} improves solubility of the additive in the Tecoflex solution and facilitates increased molecular dispersion of the additive throughout the polymer matrix.

X-ray diffraction analysis (Figure 4) confirmed crystallinity observed via DSC. Diffraction peaks corresponding to the (220), (331), (222), and (420) peaks of crystallized C_{60} were observed in the Tecoflex films containing unmodified C_{60} additive. The intensity of diffraction peaks increased with increased loading of C_{60} from 0.5 to 5.0 wt%. In contrast, control Tecoflex and Tecoflex loaded with EO_3–C_{60} displayed only a broad peak resulting from the amorphous polymer. Therefore, C_{60} aggregates into crystals in Tecoflex matrix, while EO_3–C_{60} is well dispersed.

Figure 3. Comparison of DSC thermograms for Tecoflex films containing C_{60} and neat C_{60}.

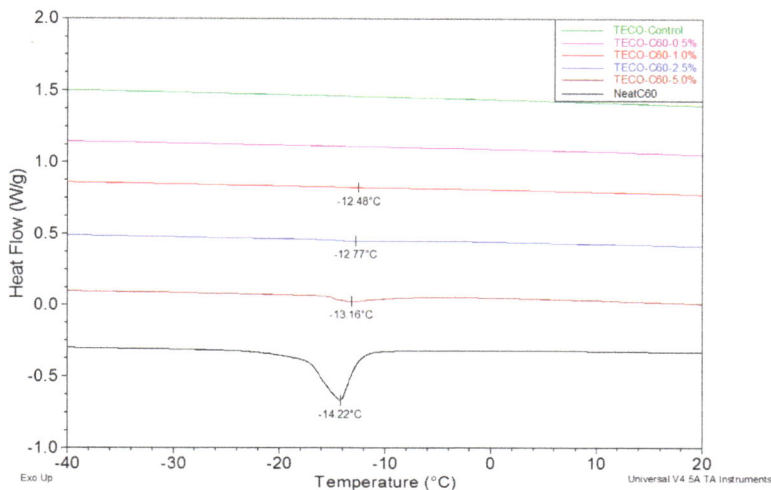

Figure 4. XRD patterns of Tecoflex films containing additives.

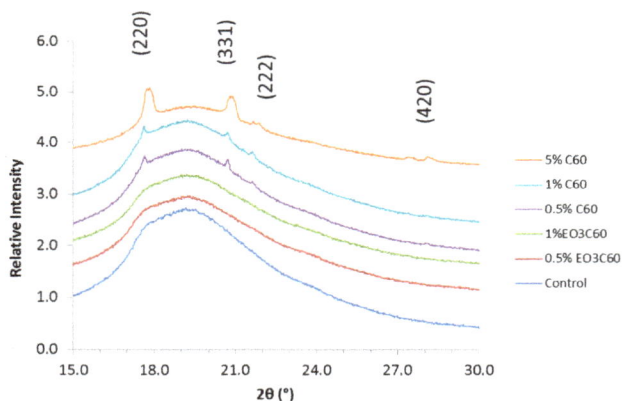

Water contact angle measurements were performed on Tecoflex films containing C_{60} and EO_3–C_{60}, the results of which are shown in Figure 5. Addition of C_{60} in Tecoflex resulted in minor increases in water contact angle in loadings up to 2.5 wt% and ultimately a minor decrease at 5 wt% C_{60} loading. Correspondingly, the surface roughness of Tecoflex films increased with increased C_{60} loading. On the other hand, loading of EO_3–C_{60} in Tecoflex resulted in a significant decrease in water contact angle accompanied with a linear increase surface roughness to a greater degree than C_{60}. Therefore, comparison of contact angle and surface roughness indicates that surface roughness plays an insignificant role in the water contact angle, despite previous evidence to the contrary [39]. Thus the affect that additive loading has on water contact angle must result from the additive's effect on the surface energy of Tecoflex, instead of imparted surface roughness.

Figure 5. Water contact angle (**a**) and surface roughness (**b**), Sq, of Tecoflex films loaded with C_{60} and EO_3–C_{60}.

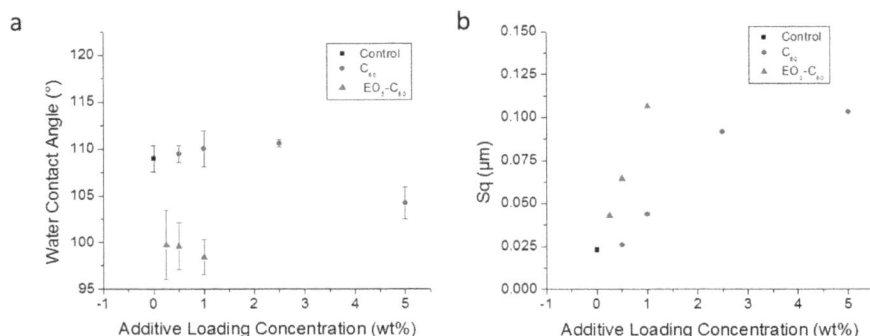

The covalent attachment of ethylene glycol moiety to C_{60} results in a molecule with amphiphilic character, *i.e.*, a surfactant. When incorporated into a solution of hydrophobic Tecoflex, the amphiphilic EO_3–C_{60} has the potential to orient its hydrophilic moiety at the polymer–air interface in order to minimize solvophobic interactions between hydrophilic ethylene oxide and hydrophobic Tecoflex matrix. Indeed, a linear decrease in contact angle was observed between 0.25 and 1.0 wt% loadings of EO_3–C_{60}. Considering that the highest loading (5%) of C_{60} in Tecoflex resulted in decrease in water contact angle of only 4°, the significant decrease in water contact angle (increase in hydrophilicity) of 10° at only 1 wt% loading of EO_3–C_{60} indicates that the additive is concentrated at the surface of the polyurethane film.

3.2. Decontamination Challenges

Decontamination challenges were employed to investigate the capability imparted by photoactive additives C_{60} and EO_3–C_{60} onto Tecoflex films to automatically decompose surface-residing contaminants. Two chemical compounds, Demeton-*S* and CEPS, were employed as representative contaminants of organophosphorous and sulfide-based pesticides. The prepared films were first subjected to decontamination challenges consisting of 18 h contaminant residence time.

Figure 6 presents results from a 4 g/m² Demeton-*S* decontamination challenge in which extracted Demeton-*S* (normalized by the tetralin internal standard) is plotted against additive loading concentration. No correlation was observed between C_{60} loading concentration in Tecoflex films and Demeton-*S* reduction. Furthermore, reduction of Demeton-*S* on C_{60} exhibits similar trends in both dark and light conditions. Tecoflex films containing EO_3–C_{60} exhibit increased reduction of Demeton-*S* compared to films containing C_{60} in dark conditions; however, decomposition of Demeton-*S* did not directly correlate with additive loading. Tecoflex films exposed to light that contained EO_3–C_{60} exhibit direct correlation between reduction of Demeton-*S* and EO_3–C_{60} loading concentration. From this, it is proposed that different modes of action for Demeton-*S* degradation are occurring between Tecoflex films containing C_{60} and EO_3–C_{60}, thus necessitating decomposition byproducts analysis.

Figure 6. Demeton-*S* recovered from Tecoflex films after 18 h residence in dark (**a**) and daylight (**b**) conditions.

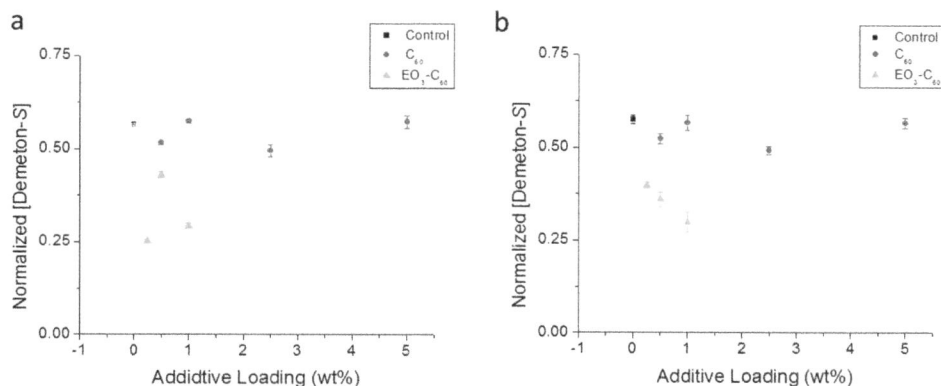

Byproduct analysis of the 18 h Demeton-*S* decontamination challenges were performed for further insight into possible modes of action. In addition to reduction in Demeton-*S*, significant byproducts were detected at a retention time of 4.1 min which corresponds to vinyl oxidation product (*S*-vinyl) (**2**). Byproduct concentration, normalized with the internal standard, is plotted against loading concentration in Figure 7. It is apparent that increased C_{60} loading leads to decreased production of **2** in both dark and light conditions, whereas increasing concentration of EO_3–C_{60} loading results in increasing production of **2**. Unmodified C_{60} is more reactive than EO_3–C_{60}, especially at low concentrations; however, increased concentration of C_{60} does not result in increased decomposition, most likely due to self-quenching resulting from high C_{60} proximity from aggregation in the polymer matrix. Qualitative observation indicated that poor solubility of C_{60} in chloroform facilitates the formation of C_{60} aggregates, in which the probability of self-quenching between C_{60} molecules is increased.

For Tecoflex films containing C_{60}, comparable amounts of **2** were detected from decomposition of Demeton-*S* on films that resided in dark and light conditions for 18 h (Figure 7a). From this, it appears that photoactivity against Demeton-*S* is not occurring in the Tecoflex films containing unmodified C_{60}. If photoactivity was occurring, then a greater amount of **2** would be observed on the films exposed to light than in darkness. It has been previously demonstrated that C_{60} fullerene can behave as an electron acceptor (Lewis acid) toward sulfides [40]. Thus, the electron acceptor behavior of C_{60} in Tecoflex may facilitate the cleavage of the S–C bond (Figure 8) resulting in the elimination product **2**.

Figure 7. Detected *S*-vinyl product (shown as normalized ratio of *S*-vinyl peak area to tetralin peak area) from Tecoflex films loaded with C_{60} (**a**) and EO_3–C_{60} (**b**) over an 18 h Demeton-*S* challenge.

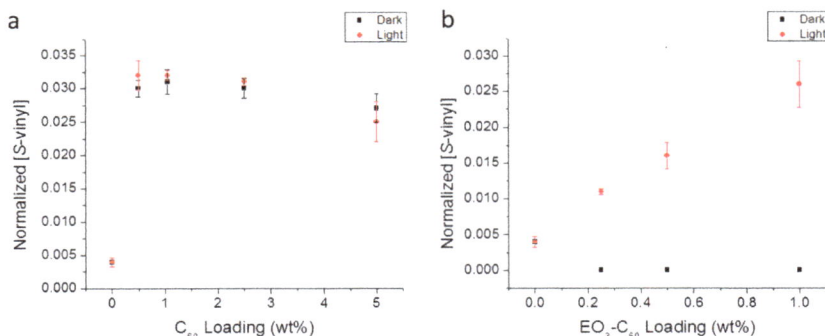

Figure 8. Hypothesized Lewis acid catalyzed sulfide elimination.

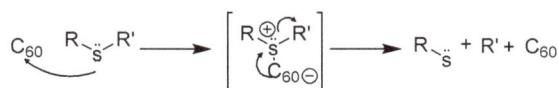

On the other hand, photoactivity was apparent from byproduct analysis in the films loaded with EO_3–C_{60} (Figure 7b). In fact, the photoactivity of EO_3–C_{60} continued to increase with increasing concentration indicating that the ethylene oxide moieties, by improving solubility, help to diminish aggregate facilitated self-quenching. In contrast to the films containing C_{60}, an absence of Demeton-*S* decomposition on Tecoflex films containing EO_3–C_{60} in dark conditions was observed. This is likely the result of decreased Lewis acid character upon the covalent attachment of the ethylene oxide moiety.

The combination of photoactivity and oxidation products detected from Tecoflex films containing EO_3–C_{60}, and documented capability of fullerene species to photogenerate singlet oxygen [41] has led to the proposed mode of action for Demeton-*S* decomposition on films containing EO_3–C_{60} exposed to visible light presented in Figure 9. The photoactive species embedded in the polymer matrix is first photosensitized upon the absorption of visible light. Subsequent transfer of energy from photosensitized EO_3–C_{60} to ambient atmospheric oxygen results in the formation of singlet oxygen ($^1O_2$). The photogenerated singlet oxygen, a ROS, then oxidizes the peripheral sulfur of Demeton-*S* that is residing on the coating surface in proximity to the photosensitized additive. The sulfoxide then undergoes α elimination resulting in the Demeton-*S* vinyl degradation product (**2**) and unstable sulfenic acid, which quickly self-condenses to form the corresponding thiosulfanate.

In addition to decontamination challenges against Demeton-*S*, Tecoflex films loaded with photoactives were also subjected to CEPS decontamination challenges. An 18 h decontamination challenge was initially performed for each sample in both dark and light conditions. Despite minor differences in the amount of CEPS decomposed, significant differences in byproduct formation were observed to be dependent on the conditions in which the sample resided.

Figure 9. Proposed oxidation mechanism of Demeton-S and the formation of elimination product from photogenerated singlet oxygen.

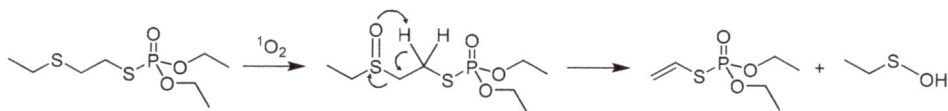

GC-MS analysis afforded the detection of a notable degradation product of CEPS from the Tecoflex films containing photoactives that resided in daylight conditions. Mass spectra analysis determined that the degradation product was vinyl phenyl sulfoxide (**4**), an oxidation byproduct of CEPS (Figure 1). Furthermore, **4** was not detected from coatings that resided in dark conditions. Figure 10 presents normalized concentrations of **4** detected in the reaction extract from Tecoflex films loaded with C_{60} and EO_3–C_{60} after an 18 h residence time of CEPS. Production of **4** from Tecoflex films decreases with increasing C_{60} loading concentration. This is attributed to increasing C_{60} aggregation with increased C_{60} loading concentration due to poor solubility and incompatibility with the polyurethane matrix, as previously demonstrated via DSC and XRD. The increase in aggregation effectively limits the available surface area of C_{60} available for both singlet oxygen generation and contact with the contaminant. Additionally, singlet oxygen quenching is known to occur between proximal C_{60} molecules in high concentration, such as in aggregates and crystallites [42].

In contrast to the effects observed resulting from C_{60} loading concentration, the generation of **4** increased with increasing EO_3–C_{60} concentration in Tecoflex from 0.25 to 1.0 wt%. This direct correlation of EO_3–C_{60} loading and generation of **4** can only result from minimized self-quenching due to good dispersion of EO_3–C_{60}. These trends support those that were observed in the Demeton-S decontamination challenge.

Figure 10. CEPS byproduct resulting from residence on Tecoflex films of several additive concentrations following exposure to daylight conditions for 18 h.

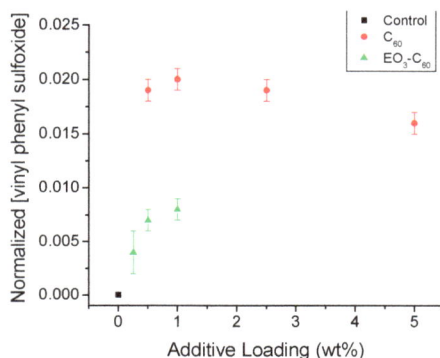

From the above 18 h study, 1 wt% loadings of C_{60} and EO_3–C_{60} were down-selected for an expanded time-dependent CEPS decontamination challenge over the course of 165 h in daylight conditions (Figure 11). Figure 11a displays concentration of CEPS extracted from samples over a 165 h residence time period. Each of the films subjected to the challenge exhibited rapid decrease in

CEPS concentration over the first 48 h. This is attributed to inherent attenuation of CEPS as this behavior was observed on the two controls, a Teflon film and unmodified Tecoflex. C_{60} and EO_3–C_{60} films differentiate from the controls at time points beyond 48 h, after which the degradation rates of CEPS on the controls decrease drastically. In contrast, the C_{60} and EO_3–C_{60} Tecoflex films exhibit continued linear degradation of CEPS beyond 48 h. This behavior is explained by two separate degradation mechanisms occurring simultaneously. First, the attenuation mechanism, which is initially at a high rate, dominates the first 48 h of degradation. Beyond 48 h, attenuation rate slows to an extent such that the secondary mechanism, photo-oxidation, dominates the overall reaction rate and thus becomes apparent.

Byproduct analysis (Figure 11b) confirms linear increase of oxidation product (4) over time on films containing photoactive additives. Tecoflex films containing C_{60} and EO_3–C_{60} each exhibited a linear relationship between production of oxidation product and residence time, while oxidation was not detected from the controls. The detection of oxidation products only from samples that are exposed to light, that contain either C_{60} or EO_3–C_{60}, and dependent on the concentration of C_{60} and EO_3–C_{60} indicates that a photo-mediated oxidation process is facilitating the oxidation of CEPS.

In consideration of the data presented herein and the known singlet oxygen generation potential of C_{60} fullerene, a proposed mode of action for CEPS oxidation is presented in Figure 12. Similar to the mechanism proposed for the oxidation of Demeton-S, the photoactive in Tecoflex generates singlet oxygen upon exposure to light. Singlet oxygen oxidizes the S atom to sulfoxide, which then through an elimination mechanism produces a vinyl sulfoxide with HCl as a byproduct. In contrast to the mechanism proposed for Demeton-S, the oxygen remains on the same byproduct molecule as sulfur. This is due to the stability of the sulfenic acid leaving group in Demeton-S decomposition compared to the instability of the potential hypochlorite byproduct in CEPS oxidation.

Figure 11. Concentration of CEPS recovered from samples over a 165 h residence (**a**) and normalized vinyl phenyl sulfoxide degradation product (**b**).

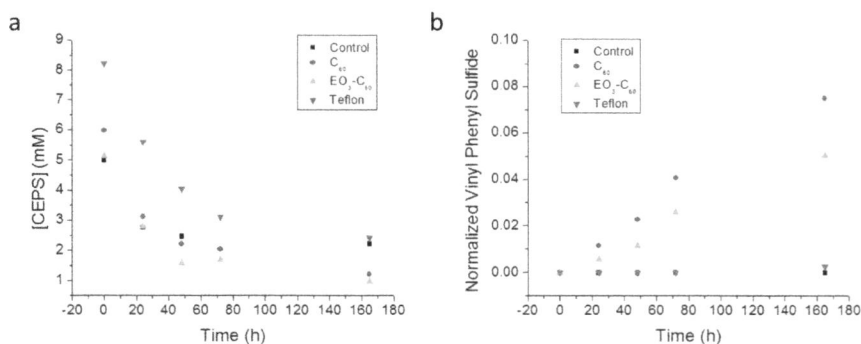

Figure 12. Potential pathway of sulfide oxidation followed by dehydrohalogenation of CEPS.

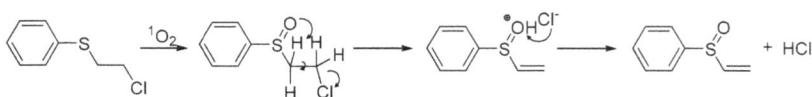

4. Conclusions

Several insights were gained upon the incorporation of C_{60} and EO_3–C_{60} into Tecoflex films. The detection of the phase transition from simple cubic to face-centered cubic structure at approximately -14 °C, subsequent confirmation with XRD, and the decreased reactivity against simulants at increased C_{60} loading concentrations provide evidence of C_{60} aggregation in Tecoflex films. Conversely, no such evidence was observed to indicate that EO_3–C_{60} was aggregating in Tecoflex films. Therefore, the covalent attachment of ethylene glycol tail to C_{60} results in an additive that exhibits improved solubility and dispersion in the hydrophobic polyurethane matrix of Tecoflex.

Surface decontamination challenges demonstrated the self-cleaning capability of Tecoflex films containing C_{60} and EO_3–C_{60} additives. Decomposition products resulting from oxidation were observed in addition to a direct correlation between additive loading concentration and decomposition of surface-residing contaminants. Through correlation of trends observed from Demeton-S and CEPS decontamination challenges and byproduct analysis, modes of action were proposed for the decomposition pathways of both contaminants on the surface of polyurethane films.

Acknowledgments

This work was funded by the Office of Naval Research (ONR) and the Naval Research Laboratory.

Author Contributions

Jeffrey G. Lundin prepared polymer films, acquired and analyzed various data, and composed the manuscript. Spencer L. Giles performed XRD data acquisition and analysis. Robert F. Cozzens assisted in interpretation of degradation reactions. James H. Wynne assisted in manuscript revision and project management.

Conflicts of Interest

The authors declare no conflict of interest.

References

1. Bartelt-Hunt, S.L.; Knappe, D.R.U.; Barlaz, M.A. A review of chemical warfare agent simulants for the study of environmental behavior. *Crit. Rev. Environ. Sci. Technol.* **2008**, *38*, 112–136.
2. Gephart, R.T.; Coneski, P.N.; Wynne, J.H. Decontamination of chemical-warfare agent simulants by polymer surfaces doped with the singlet oxygen generator zinc octaphenoxyphthalocyanine. *ACS Appl. Mater. Interfaces* **2013**, *5*, 10191–10200.
3. Chattopadhyay, D.K.; Raju, K.V.S.N. Structural engineering of polyurethane coatings for high performance applications. *Prog. Polym. Sci.* **2007**, *32*, 352–418.
4. Wynne, J.H.; Pant, R.R.; Jones-Meehan, J.M.; Phillips, J.P. Preparation and evaluation of nonionic amphiphilic phenolic biocides in urethane hydrogels. *J. Appl. Polym. Sci.* **2008**, *107*, 2089–2094.

5. Pant, R.R.; Buckley, J.L.; Fulmer, P.A.; Wynne, J.H.; McCluskey, D.M.; Phillips, J.P. Hybrid siloxane epoxy coatings containing quaternary ammonium moieties. *J. Appl. Polym. Sci.* **2008**, *110*, 3080–3086.

6. Kurt, P.; Wood, L.; Ohman, D.E.; Wynne, K.J. Highly effective contact antimicrobial surfaces via polymer surface modifiers. *Langmuir* **2007**, *23*, 4719–4723.

7. Harney, M.B.; Pant, R.R.; Fulmer, P.A.; Wynne, J.H. Surface self-concentrating amphiphilic quaternary ammonium biocides as coating additives. *ACS Appl. Mater. Interfaces* **2009**, *1*, 39–41.

8. Pant, R.R.; Fulmer, P.A.; Harney, M.B.; Buckley, J.P.; Wynne, J.H. Synthesis and biocidal efficacy of self-spreading polydimethylsiloxane oligomers possessing oxyethylene-functionalized quaternary ammoniums. *J. Appl. Polym. Sci.* **2009**, *113*, 2397–2403.

9. Pant, R.R.; Rasley, B.T.; Buckley, J.P.; Lloyd, C.T.; Cozzens, R.F.; Santangelo, P.G.; Wynne, J.H. Synthesis, mobility study and antimicrobial evaluation or novel self-spreading ionic silicone oligomers. *J. Appl. Polym. Sci.* **2007**, *104*, 2954–2964.

10. Fulmer, P.A.; Lundin, J.G.; Wynne, J.H. Development of antimicrobial peptides (amps) for use in self-decontaminating coatings. *ACS Appl. Mater. Interfaces* **2010**, *2*, 1266–1270.

11. Talmage, S.S.; Watson, A.P.; Hauschild, V.; Munro, N.B.; King, J. Chemical warfare agent degradation and decontamination. *Curr. Org. Chem.* **2007**, *11*, 285–298.

12. Arbogast, J.W.; Darmanyan, A.P.; Foote, C.S.; Diederich, F.N.; Whetten, R.L.; Rubin, Y.; Alvarez, M.M.; Anz, S.J. Photophysical properties of sixty atom carbon molecule (C_{60}). *J. Phys. Chem.* **1991**, *95*, 11–12.

13. Nagano, T.; Arakane, K.; Ryu, A.; Masunaga, T.; Shinmoto, K.; Mashiko, S.; Hirobe, M. Comparison of singlet oxygen production efficiency of C_{60} with other photosensitizers, based on 1268 nm emission. *Chem. Pharm. Bull.* **1994**, *42*, 2291–2294.

14. Pichler, K.; Graham, S.; Gelsen, O.M.; Friend, R.H.; Romanow, W.J.; McCauley, J.P., Jr.; Coustel, N.; Fischer, J.E.; Smith, A.B. Photophysical properties of solid films of fullerene, C_{60}. *J. Phys. Condens. Matter* **1991**, *3*, 9259–9270.

15. Wasielewski, M.R.; O'Neil, M.P.; Lykke, K.R.; Pellin, M.J.; Gruen, D.M. Triplet states of fullerenes C_{60} and C_{70}. Electron paramagnetic resonance spectra, photophysics, and electronic structures. *J. Am. Chem. Soc.* **1991**, *113*, 2774–2776.

16. Hotze, E.M.; Labille, J.; Alvarez, P.; Wiesner, M.R. Mechanisms of photochemistry and reactive oxygen production by fullerene suspensions in water. *Environ. Sci. Technol.* **2008**, *42*, 4175–4180.

17. Haufler, R.E.; Wang, L.-S.; Chibante, L.P.F.; Jin, C.; Conceicao, J.; Chai, Y.; Smalley, R.E. Fullerene triplet state production and decay: R2PI probes of C_{60} and C_{70} in a supersonic beam. *Chem. Phys. Lett.* **1991**, *179*, 449–454.

18. Fraelich, M.R.; Weisman, R.B. Triplet states of fullerene C_{60} and C_{70} in solution: Long intrinsic lifetimes and energy pooling. *J. Phys. Chem.* **1993**, *97*, 11145–11147.

19. Orfanopoulos, M.; Kambourakis, S. Chemical evidence of singlet oxygen production from C_{60} and C_{70} in aqueous and other polar media. *Tetrahedron Lett.* **1995**, *36*, 435–438.

20. Lee, J.; Fortner, J.D.; Hughes, J.B.; Kim, J.-H. Photochemical production of reactive oxygen species by C_{60} in the aqueous phase during uv irradiation. *Environ. Sci. Technol.* **2007**, *41*, 2529–2535.

21. Lee, J.; Mackeyev, Y.; Cho, M.; Li, D.; Kim, J.-H.; Wilson, L.J.; Alvarez, P.J.J. Photochemical and antimicrobial properties of novel C_{60} derivatives in aqueous systems. *Environ. Sci. Technol.* **2009**, *43*, 6604–6610.

22. Lee, J.; Yamakoshi, Y.; Hughes, J.B.; Kim, J.-H. Mechanism of C_{60} photoreactivity in water: Fate of triplet state and radical anion and production of reactive oxygen species. *Environ. Sci. Technol.* **2008**, *42*, 3459–3464.

23. Deguchi, S.; Alargova, R.G.; Tsujii, K. Stable dispersions of fullerenes, C_{60} and C_{70}, in water. Preparation and characterization. *Langmuir* **2001**, *17*, 6013–6017.

24. Lyon, D.Y.; Adams, L.K.; Falkner, J.C.; Alvarez, P.J.J. Antibacterial activity of fullerene water suspensions: Effects of preparation method and particle size. *Environ. Sci. Technol.* **2006**, *40*, 4360–4366.

25. Wilson, M. Light-activated antimicrobial coating for the continuous disinfection of surfaces. *Infect. Control Hosp. Epidemiol.* **2003**, *24*, 782–784.

26. Page, K.; Wilson, M.; Parkin, I.P. Antimicrobial surfaces and their potential in reducing the role of the inanimate environment in the incidence of hospital-acquired infections. *J. Mater. Chem.* **2009**, *19*, 3818–3831.

27. Perni, S.; Piccirillo, C.; Pratten, J.; Prokopovich, P.; Chrzanowski, W.; Parkin, I.P.; Wilson, M. The antimicrobial properties of light-activated polymers containing methylene blue and gold nanoparticles. *Biomaterials* **2009**, *30*, 89–93.

28. Belousova, I.M.; Danilov, O.B.; Muraveva, T.D.; Kiselyakov, I.M.; Rylkov, V.V.; Krisko, T.K.; Kiselev, O.I.; Zarubaev, V.V.; Sirotkin, A.K.; Piotrovskii, L.B. Solid-phase photosensitizers based on fullerene C_{60} for photodynamic inactivation of viruses in biological liquids. *J. Opt. Technol.* **2009**, *76*, 243–250.

29. McCluskey, D.M.; Smith, T.N.; Madasu, P.K.; Coumbe, C.E.; Mackey, M.A.; Fulmer, P.A.; Wynne, J.H.; Stevenson, S.; Phillips, J.P. Evidence for singlet-oxygen generation and biocidal activity in photoresponsive metallic nitride fullerene-polymer adhesive films. *ACS Appl. Mater. Interfaces* **2009**, *1*, 882–887.

30. Vandewal, K.; Albrecht, S.; Hoke, E.T.; Graham, K.R.; Widmer, J.; Douglas, J.D.; Schubert, M.; Mateker, W.R.; Bloking, J.T.; Burkhard, G.F.; *et al.* Efficient charge generation by relaxed charge-transfer states at organic interfaces. *Nat. Mater.* **2014**, *13*, 63–68.

31. Badamshina, E.; Estrin, Y.; Gafurova, M. Nanocomposites based on polyurethanes and carbon nanoparticles: Preparation, properties and application. *J. Mater. Chem. A* **2013**, *1*, 6509–6529.

32. Giacalone, F.; Martín, N. Fullerene polymers: Synthesis and properties. *Chem. Rev.* **2006**, *106*, 5136–5190.

33. Accorsi, G.; Armaroli, N. Taking advantage of the electronic excited states of [60]-fullerenes. *J. Phys. Chem. C* **2010**, *114*, 1385–1403.

34. Badamshina, E.; Gafurova, M. Polymeric nanocomposites containing non-covalently bonded fullerene C_{60}: Properties and applications. *J. Mater. Chem.* **2012**, *22*, 9427–9438.

35. Hawker, C.J.; Saville, P.M.; White, J.W. The synthesis and characterization of a self-assembling amphiphilic fullerene. *J. Org. Chem.* **1994**, *59*, 3503–3505.

36. Teh, S.-L.; Linton, D.; Sumpter, B.; Dadmun, M.D. Controlling non-covalent interactions to modulate the dispersion of fullerenes in polymer nanocomposites. *Macromolecules* **2011**, *44*, 7737–7745.

37. Sachidanandam, R.; Harris, A.B. Comment on "orientational ordering transition in solid C_{60}". *Phys. Rev. Lett.* **1991**, *67*, 1467–1467.

38. Heiney, P.A.; Fischer, J.E.; McGhie, A.R.; Romanow, W.J.; Denenstein, A.M.; McCauley, J.P., Jr.; Smith, A.B.; Cox, D.E. Orientational ordering transition in solid C_{60}. *Phys. Rev. Lett.* **1991**, *66*, 2911–2914.

39. Bellanger, H.; Darmanin, T.; Taffin de Givenchy, E.; Guittard, F. Chemical and physical pathways for the preparation of superoleophobic surfaces and related wetting theories. *Chem. Rev.* **2014**, *114*, 2694–2716.

40. Zhang, J.Z.; Geselbracht, M.J.; Ellis, A.B. Binding of fullerenes to cadmium sulfide and cadmium selenide surfaces, photoluminescence as a probe of strong, lewis acidity-driven, surface adduct formation. *J. Am. Chem. Soc.* **1993**, *115*, 7789–7793.

41. Phillips, J.P.; Deng, X.; Todd, M.L.; Heaps, D.T.; Stevenson, S.; Zhou, H.; Hoyle, C.E. Singlet oxygen generation and adhesive loss in stimuli-responsive, fullerene-polymer blends, containing polystyrene-block-polybutadiene-block-polystyrene and polystyrene-block-polyisoprene-block-polystyrene rubber-based adhesives. *J. Appl. Polym. Sci.* **2008**, *109*, 2895–2904.

42. Bagrov, I.; Belousova, I.; Danilov, O.; Kiselev, V.; Murav'eva, T.; Sosnov, E. Photoinduced quenching of the luminescence of singlet oxygen in fullerene solutions. *Opt. Spectrosc.* **2007**, *102*, 52–59.

Chapter 3:
Photocatalytic Removal of Air Pollutants

Recent Photocatalytic Applications for Air Purification in Belgium

Elia Boonen and Anne Beeldens

Abstract: Photocatalytic concrete constitutes a promising technique to reduce a number of air contaminants such as NO_x and VOC's, especially at sites with a high level of pollution: highly trafficked canyon streets, road tunnels, the urban environment, *etc.* Ideally, the photocatalyst, titanium dioxide, is introduced in the top layer of the concrete pavement for best results. In addition, the combination of TiO_2 with cement-based products offers some synergistic advantages, as the reaction products can be adsorbed at the surface and subsequently be washed away by rain. A first application has been studied by the Belgian Road Research Center (BRRC) on the side roads of a main entrance axis in Antwerp with the installation of 10.000 m² of photocatalytic concrete paving blocks. For now however, the translation of laboratory testing towards results *in situ* remains critical of demonstrating the effectiveness in large scale applications. Moreover, the durability of the air cleaning characteristic with time remains challenging for application in concrete roads. From this perspective, several new trial applications have been initiated in Belgium in recent years to assess the "real life" behavior, including a field site set up in the Leopold II tunnel of Brussels and the construction of new photocatalytic pavements on industrial zones in the cities of Wijnegem and Lier (province of Antwerp). This paper first gives a short overview of the photocatalytic principle applied in concrete, to continue with some main results of the laboratory research recognizing the important parameters that come into play. In addition, some of the methods and results, obtained for the existing application in Antwerp (2005) and during the implementation of the new realizations in Wijnegem and Lier (2010–2012) and in Brussels (2011–2013), will be presented.

Reprinted from *Coatings*. Cite as: Boonen, E.; Beeldens, A. Recent Photocatalytic Applications for Air Purification in Belgium. *Coatings* **2014**, *4*, 553-573.

1. Introduction

Emission from the transport sector has a particular impact on the overall air quality because of its rapid rate of growth: goods transport by road in Europe (EU-27) has increased by 31% (period 1995–2009), while passenger transport by road in the EU-27 has gone up by 21% and passenger transport in air by 51% in the same period [1]. The main emissions caused by motor traffic are nitrogen oxides (NO_x), hydrocarbons (HC) and carbon monoxide (CO), accounting for respectively 46%, 50% and 36% of all such emissions in Europe in 2008 [2].

These pollutants have an increasing impact on the urban air quality. In addition, photochemical reactions resulting from the action of sunlight on NO_2 and VOC's (volatile organic compounds) lead to the formation of "photochemical smog" and ozone, a secondary long-range pollutant, which impacts in rural areas often far from the original emission site. Acid rain is another long-range pollutant influenced by vehicle NO_x emissions and resulting from the transport of NO_x, oxidation in

178

the air into HNO$_3$ and finally, precipitation of (acid) NO$_3^-$ with harmful consequences for building materials (corrosion of the surface) and vegetation.

The European Directives [3] impose a limit to the NO$_2$ concentration in ambient air of maximum 40 μg/m^3 NO$_2$ (21 ppbV) averaged over 1 year and 200 μg/m^3 (106 ppbV) averaged over 1 h. These limit values gradually decreased from 50 and 250 in 2005 to the final limit in 2010.

Heterogeneous photocatalysis is a promising method for NO$_x$ abatement. In the presence of UV-light, the photocatalytically active form of TiO$_2$ present at the surface of the material is activated, enabling the abatement of pollutants in the air. The translation from laboratory results to real cases is starting. Different applications are implemented in Belgium in order to see the influence of the photocatalytic materials on real scale and to determine the durability of the air purifying capacity over time.

In the first part of this paper, the principle of photocatalytic concrete will be elaborated, followed by a description of the past laboratory research indicating important influencing factors for the purifying process. Next, an overview of the results regarding the first pilot project in Antwerp [2] is given, and finally, the different applications in Belgium that have recently been finished, will be discussed.

2. Photocatalytic Concrete: Purifying the Air through the Pavement

A solution for the air pollution by traffic can be found in the treatment of the pollutants as close to the source as possible. Therefore, photocatalytically active materials can be added to the surface of pavement and building materials [4]. Air purification through heterogeneous photocatalysis consists of different steps: under the influence of UV-light, the photoactive TiO$_2$ at the surface of the material is activated. Subsequently, the pollutants are oxidized due to the presence of the photocatalyst and precipitated on the surface of the material. Finally, they can be removed from the surface by the rain or cleaning/washing with water, see Figure 1.

Figure 1. Schematic of photocatalytic air purifying pavement.

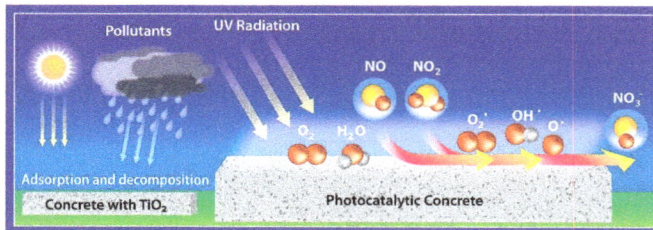

Heterogeneous photocatalysis with titanium dioxide (TiO$_2$) as catalyst is a rapidly developing field in environmental engineering, as it has a great potential to cope with the increasing pollution. Besides its self-cleaning properties, it is known since almost 100 years that titanium dioxide acts as a photo-catalyst that can decompose pollutants under UV radiation [5]. The impulse for the more widespread use of TiO$_2$ photocatalytic materials was further given in 1972 by Fujishima and Honda [6], who discovered the hydrolysis of water in the presence of light, by means of a TiO$_2$-anode in a photochemical cell. In the 1980s, organic pollution in water was also decomposed by adding TiO$_2$

and under influence of UV-light [7]. The application of TiO_2, in the photo-active crystalline phase anatase, as air purifying material originated in Japan in 1996 (e.g., [8]). Since then, a broad spectrum of products appeared on the market for indoor use as well as for outdoor applications. Regarding traffic emissions, it is important that the exhaust gases stay in contact with the active surface during a certain period. The street configuration, the speed of the traffic, the speed and direction of the wind, all influence the final reduction rate of pollutants *in situ*.

In the case of concrete pavement blocks [9,10], the anatase is added to the wearing layer of the pavers which is approximately 8 mm thick. In the case of cast-in-place concrete pavements, the TiO_2 is added in the top layer (40 mm thick). The fact that the TiO_2 is present over the whole thickness of this layer means that even if some surface wear takes place, for example by traffic or weathering, new TiO_2 will be present at the surface to maintain the photocatalytic activity (in contrast to the abrasion of a coating or dispersion layer for instance). The use of TiO_2 in combination with cement leads to a transformation of the NO_x into NO_3^-, which is adsorbed at the surface due to the alkalinity of the concrete [11]. Thus, a synergetic effect is created in the presence of the cement matrix, which helps to effectively trap the reactant gases (NO and NO_2) together with the nitrate salt formed. Subsequently, the deposited nitrate can be washed away by rain or washing with water. In addition, these nitrates pose no real threat towards pollution of body waters because the resulting concentrations in the waste water are very low, even below the current limit values for surface and ground water [12].

Special attention is given here to the NO and NO_2 content in the air, since they are for almost 50% caused by the exhaust of traffic and are at the base of smog, secondary ozone and acid rain formation as indicated above. The photocatalytic oxidation of NO is usually assumed to be a surface reaction between NO and an oxidizing species formed upon the adsorption of a photon by the photocatalyst, e.g., a hydroxyl radical, both adsorbed at the surface of the photocatalyst [13]. It has been shown by various authors that the final product of the photocatalytic oxidation of NO in the presence of TiO_2 is nitric acid (HNO_3) while HNO_2 and NO_2 have been identified as intermediate products in the gas phase over the photocatalyst [2,4,11,13,14]. The resulting reaction pathway of the photocatalytic oxidation of NO has been discussed in several publications e.g., [2,4,13–16] most of which proposed the photocatalytic conversion of NO via HNO_2 to yield NO_2, which is subsequently oxidized by the attack of a hydroxyl radical to the final product HNO_3:

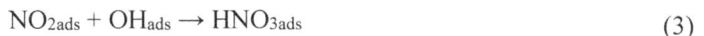

$$NO_{ads} + OH_{ads} \rightarrow HNO_{2ads} \tag{1}$$

$$HNO_{2ads} + OH_{ads} \rightarrow NO_{2ads} + H_2O_{ads} \tag{2}$$

$$NO_{2ads} + OH_{ads} \rightarrow HNO_{3ads} \tag{3}$$

Here, all nitrogen compounds adsorbed at the photocatalyst surface are assumed to be in equilibrium with the gas phase.

Until now, UV-light (in the UV-A spectrum) was necessary to activate the photocatalyst. However, recent research indicates a shift towards the visible light [17], for instance by doping the TiO_2 with transition metal ions or non-metallic anionic species, or forming reduced TiO_x. These techniques introduce impurities and defects in the band gap of TiO_2 thereby increasing the amount of visible light that can be absorbed and used in the photocatalytic process. This means that applications in tunnels and indoor environments become more realistic. Especially the application in tunnels is worth looking at due to the high concentration of air pollutants at these sites. One of the projects in Belgium is focusing on this subject [18].

3. Laboratory Results: Parameter Evaluation

Different test methods have been developed to determine the efficiency of photocatalytic materials towards air purification. An overview is given in [11]. A distinction can be made by the type of air flow; in the flow-through method according to ISO 22197-1 [19], the air, with a concentration of 1 ppmV of NO, passes once-only over the sample which is illuminated by a UV-lamp with light intensity equal to 10 W/m^2 in the range between 300 and 400 nm, as illustrated in Figure 2. Afterwards, the NO_x (= sum of NO and NO_2) concentration is measured at the outlet and the reduction (in %) is calculated. It is also worth to note here that within Europe actions are underway to harmonize and develop new standards for photocatalyis [20]. In any case, the test procedure used for the current results is still based on the existing ISO standard.

Figure 2. (**a**) Schematic and (**b**) photo of measurement set-up based on ISO 22197-1:2007 [19] at Belgian Road Research Center (BRRC).

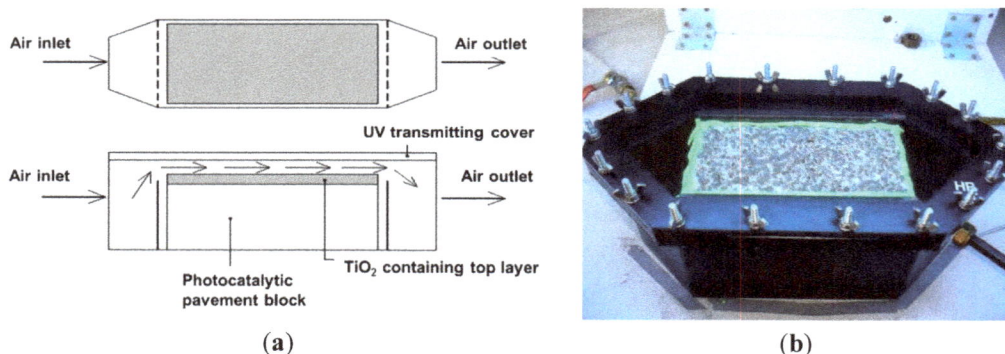

(**a**)　　　　　　　　　　　　　　　　(**b**)

The pre-treatment of the samples in the laboratory can be important to obtain reproducible results and mainly depends on the type of base material (e.g., concrete or paints). A typical test scheme according to the ISO standard is presented in Figure 3, where the following steps are applied to the sample: 0.5 h at 1 ppmV NO concentration, no light—5 h exposure to an air flow of 3 L/min with 1 ppmV NO and UV-illumination—0.5 h with UV-illumination and no exposure. A small increase with time of the NO_x concentration is visible due to the deposit of the NO_3^- at the surface.

Figure 3. Typical result obtained in the laboratory following the standard ISO test procedure.

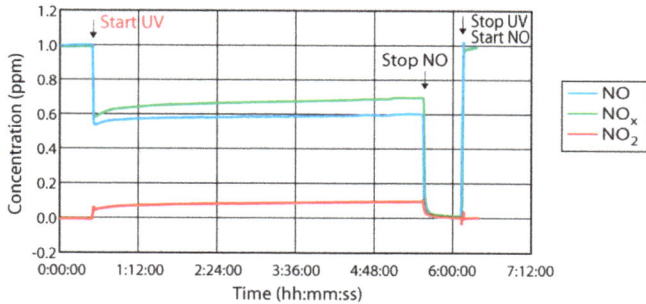

The influence of different important test parameters affecting the photocatalytic reaction has been investigated before [2] such as temperature, light intensity, relative humidity, contact time (controlled by surface area, flow velocity, height of air flow, *etc.*). For instance, the effect of relative humidity of the ingoing air is illustrated in Figure 4 for different materials including cementitious (concrete, mortar) and other (paint) substrates. Clearly, for cementitious materials the reduction of the NO_x concentration in the outlet air decreases with increasing relative humidity (RH, %), an observation which was also found by other authors [21]. This probably has to do with the fact that the water in the atmosphere plays a role in the adhesion of the pollutants at the surface and with the competition effect that can arise between water molecules and NO_x in the ambient air with increasing RH. For paints (acidic environment) though, it has been noticed that there is an optimum in RH where a maximal efficiency is obtained. Anyway, relative humidity proves to be an important limiting factor for photocatalytic applications in humid areas like Belgium. Temperature on the other hand, was found to have no significant influence on the NO_x reduction within the ambient range (5–25 °C).

Figure 4. Effect of relative humidity on photocatalytic efficiency for different materials.

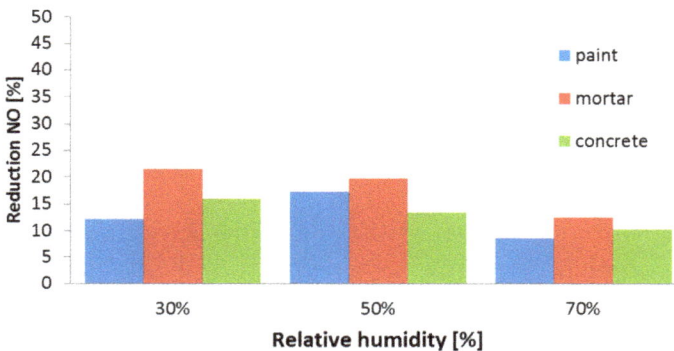

In general, it can be stated that the efficiency towards the reduction of NO_x (in %) increases with a longer contact time (larger surface area, lower air velocity, smaller height of air flow, higher turbulence at the surface), a lower relative humidity (for cementitious materials) and a higher intensity of incident light. These are the conditions at which the risk of ozone formation in summer is the largest: higher sun light intensity, no wind and no rain. At these days, the photocatalytic reaction will be more pronounced.

4. Pilot Project in Antwerp

An important issue is the conversion of the results obtained in the laboratory to real applications. In order to see the influence of photocatalytic pavements in "real conditions", a first pilot section of 10.000 m² of photocatalytic pavement blocks was constructed in 2004–2005 on the parking lanes of a main axe in Antwerp [2]. Figure 5 depicts a view of the parking lane, where the photocatalytic concrete pavement blocks have been applied. Only the wearing layer (upper 5–6 mm) of the blocks contains anatase TiO_2 mixed in the mass of the concrete layer. The exact composition could not be given by the manufacturer (Marlux, Tessenderlo, Belgium) at that time in view of confidentiality. In spite of the fact that the surface applied on the Leien of Antwerp is quite important, one has to notice the relatively small width of the photocatalytic parking lanes in comparison with the total street: 2×4.5 m *versus* a total width of 60 m.

Figure 5. Separate parking lanes at the Leien of Antwerp with photocatalytic pavement blocks.

In order to check the durability of the photocatalytic efficiency, pavement blocks were taken from the lane after different periods of exposure and measured in the laboratory with and without washing of the surface. Some of the results are presented in Figure 6. They indicate a good durability of the efficiency towards NO_x abatement. The deposition of pollutants on the surface leads to a decrease in efficiency which can be regained after washing. Recently repeated measurements in 2010 indicate that even after more than five years of service life, the photocatalytic efficiency of the pavers is still present [22].

Besides the tests in the lab, on site measurements were also carried out. Since no reference measurements without photocatalytic material (prior to the application) exist, the interpretation of these results is rather difficult. Especially the influence of traffic, wind speed, light intensity and relative humidity are playing an important role. Detailed results can be found in [2]. In brief, the field measurements suggested a decrease in NO_x concentration at the sites with photocatalytic materials, where a levelling out of the pollution peaks is visible. In any case, precaution has to be taken with the interpretation of data since these results are momentary and limited over time. However, at least, they gave an indication of the efficiency of the photocatalytic pavement materials *in situ*, and a basis to work on for future applications.

Figure 6. NO$_x$ reduction measured on two pavement blocks, before (hatched) and after (colored) washing the surface, taken on different locations (house nr. 30, 35, 37, 42, 48, 53) and at different times at the Amerikalei in Antwerp.

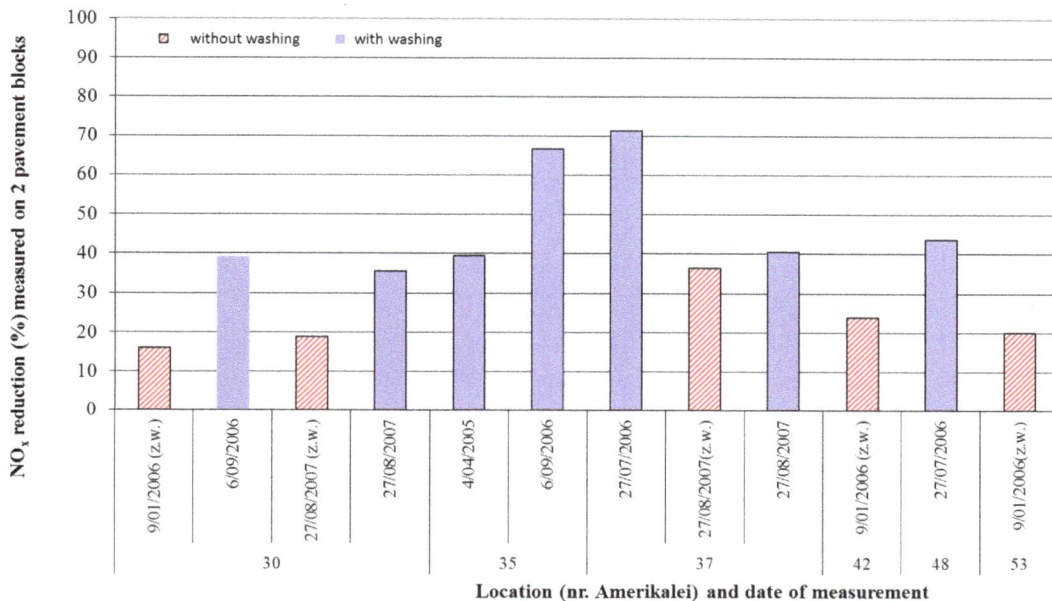

5. Recent Photocatalytic Applications in Belgium

Since the first application in Antwerp (2004–2005), much progress has been made within the photocatalytic research area. Newer, better and more efficient materials are constantly being developed, and action is more and more broadened also to visible light responsive materials [17]. This also led to new trial applications in which people have tried to establish the relation between the results in the laboratory and the real effect on site, see e.g., [23–25]. In this section an overview is given of two such recent projects in Belgium which were implemented in collaboration with the BRRC.

5.1. Life+ Project PhotoPAQ

The European Life+ funded project PhotoPAQ [18] was aimed at demonstrating the usefulness of photocatalytic construction materials for air purification purposes in an urban environment. Eight partners from five different European countries participated in the project.

In this framework, an extensive three-step field campaign was organized in the Leopold II tunnel in Brussels, from June 2011 till January 2013 [26,27]. A photocatalytic cementitious coating material (TX Active® white Skim Coat from CTG Italcementi Group) was applied on the side walls and roof (total area of about 2700 m²) of a tunnel section of about 160 m in length in one of the tunnel tubes directing to the city center. The air-purifying product was activated by a dedicated UV lighting system (including Supratec "HTC 241 R7s" light bulbs from Osram, see Figure 7). More details can be found in [26,27].

Figure 7. Application of the photocatalytic product and installation of the UV lamps in the Leopold II tunnel in Brussels, in the framework of PhotoPAQ.

Possible advantages of purifying the tunnel air may be, obviously, cleaner air to breathe, with a potentially reduced need of ventilation, but also (and maybe mainly) a reduction of the pollution impact of tunnel exhaust on the city air quality. During the field campaigns, the effect of the photocatalytic coating on the air pollution (including NO_x, VOC's, particulate matter, CO, *etc.*) inside the tunnel section was rigorously assessed.

The PhotoPAQ consortium mobilized a large panel of up-to-date instrumentations, installed in the tunnel for several weeks, aiming at characterizing the levels of pollution in this section of the Leopold II tunnel, with and without the air-purifying product (Figure 8).

Figure 8. Full characterization of the air quality inside the tunnel test section during the PhotoPAQ campaigns.

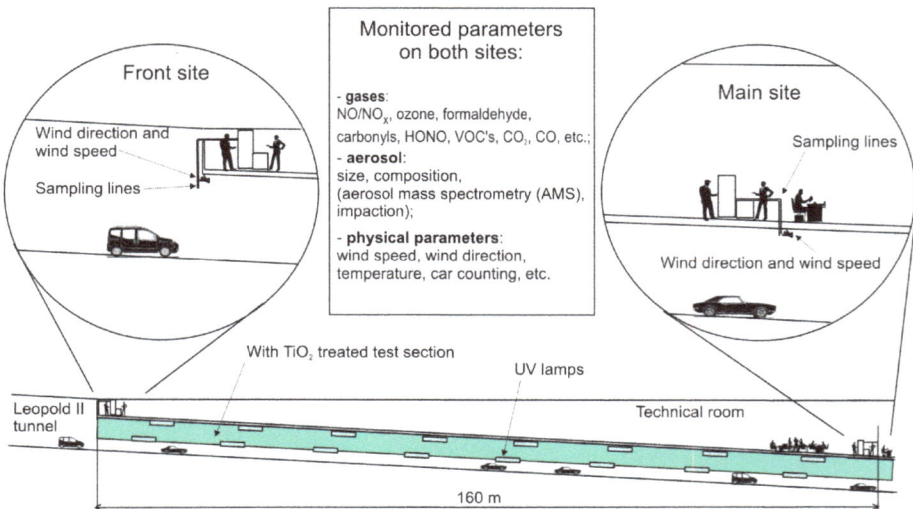

However, in contrast to first estimations based on laboratory studies, the results indicated no observable reduction of the pollution level, *i.e.*, the reduction of nitrogen oxides (NO_x, one of the major traffic related air pollutants) is below 2%, which is the experimental uncertainty of the measurements.

A severe de-activation of the photocatalytic material was observed inside the highly trafficked and strongly polluted Leopold II tunnel. In conjunction, final UV lighting intensity (only 2 W/m² UV-A) was below the targeted values (above 4 W/m²), which led to too low levels for proper activation inside the polluted tunnel environment. Another negative condition was the high wind speed (up to 3 m/s) inside the tunnel, limiting the contact time between pollutants and the active surface. Finally, January 2013 turned out to be an unusually wintry period causing cold and humid conditions inside the tunnel, with relative humidity ranging from 70% to 90%, which also reduces the activity of the photocatalytic material as shown before. Thus, all these issues together resulted in a reduction of the activity of the photocatalytic surfaces inside the harsh environment of the Leopold II tunnel, by a factor of 10 compared to the theoretical expectations. More details about the set-up and results of these extensive field campaigns inside the Leopold II tunnel are presented elsewhere [26,27].

Nevertheless, combining the knowledge gained during these campaigns and the laboratory based investigations performed by the PhotoPAQ consortium, numerical simulations (with the commercially available general purpose Computational Fluid Dynamics code ANSYS CFX®) were performed in order to estimate the possible best-case abatement of pollutants.

These calculations indicate that, under the best case scenario (proper level of UV light intensity higher than 4 W/m², relative humidity below 50%, and limited pollution to avoid passivation), the reduction of the NO_x concentration may be expected to attain:

- ±3% for the 160 m long test section;
- ±12% for the entire Leopold II tunnel (*ca.* 3 km), if not affected by ventilation.

Despite the fact that the results were not as expected, the Leopold II field campaigns conducted by the PhotoPAQ team proved to be a unique real world and fully comprehensive assessment of the effect of photocatalytic air-purifying (road) construction materials on air pollution inside a tunnel environment. Based on the extensive experimental data set gathered and numerical model calculations, a valuable tool for extrapolation can be provided to estimate the expected pollution reduction in other urban tunnel sites, also for use by non-experts [18].

5.2. INTERREG Project ECO₂PROFIT

The broad environmental sustainability project ECO₂PROFIT dealt with the reduction of the emission of greenhouse gases and sustainable production of energy on industrial estates in the frontier area between Flanders and Holland. To reach these goals, several tangible demonstration projects were carried out on industrial sites in Belgium and the Netherlands. BRRC was involved in two such projects: "Den Hoek 3" in Wijnegem and "Duwijckpark" in Lier (both near Antwerp). Here, the regional development agency POM Antwerp was aiming to use a double layered concrete for the road construction, with recycled concrete aggregates in the bottom layer and photocatalytic materials (TiO$_2$) in the top layer, using photoactive cements and/or coatings. That way, air purifying and CO$_2$ reducing concrete roads could be built which are both innovating and energy efficient.

For these recently completed applications (2010–2011) BRRC was asked to set-up an elaborate testing program in the lab to help optimize the air purifying performance of the top layer, without interfering with other properties of the concrete (workability, strength, durability *etc.*). In the construction site of Wijnegem (Den Hoek 3), it was opted to use an exposed aggregates surface finish (with grain size between 0 and 6.3 mm) on the top layer for reasons of noise reduction and comfort of the road user. For the site in Lier (Duwijckpark) a brushed surface finishing was chosen to have more active cement at the surface. Indeed, the type of surface finishing and/or treatment of the pavement can have an effect on the photocatalytic efficiency, as shown in Figure 9 for three types of surface finishing: exposed aggregates, smooth (formwork side) and sawn surface. The results show that the exposed aggregates surface performs equally well as the smooth, formwork surface, but not as good as a sawn surface. This is the result of the combined action of less photoactive cement at the surface and a higher surface porosity (higher specific surface), two competing effects which in the end yield the final efficiency shown in Figure 9.

Figure 9. Effect of surface treatment on photocatalytic efficiency (only one type of "less" active product in mass).

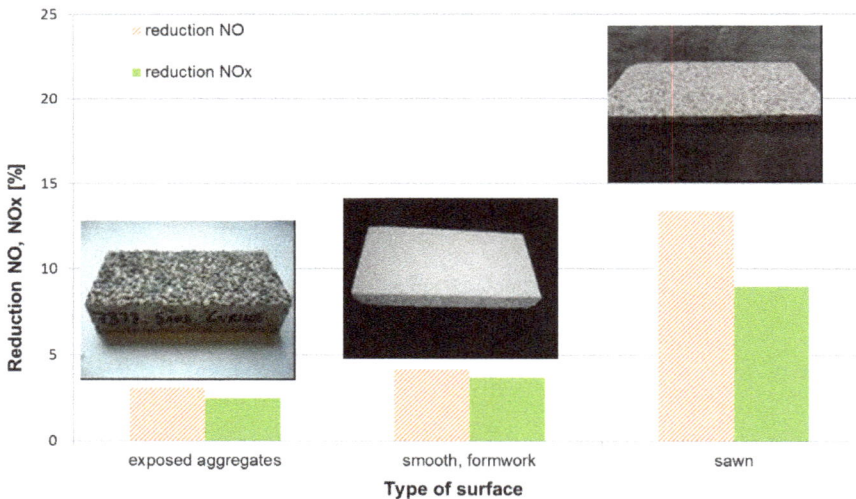

For the application of photocatalytic materials in a concrete road (and in general for any other type of application) a fundamental choice can be made between: mixing in the mass (e.g., TiO$_2$ in cement) and/or spraying on the surface (suspension of TiO$_2$). The former has the advantage of a more durable action since the TiO$_2$ will continuously be present, even after wearing of the top layer. On the other hand, the initial cost will be higher (higher TiO$_2$ content, necessity for double layered concrete) and only the TiO$_2$ at the surface will be active. In contrast, dispersing at the surface of a TiO$_2$ solution will provide a more direct action, and a lower initial cost (e.g., "ordinary" cement). In this case however, the longevity of the photocatalytic action could be questioned because of loss of adhesion to the surface in time. This fundamental choice was also investigated within the research program, together with the influence of several other parameters [28].

The effect of a curing compound for instance—generally applied to protect the young concrete against desiccation in Belgium and placed directly after concreting or after exposing the aggregates at the surface in case of denudation—is illustrated in Figure 10. From this, it appears the curing compound will initially inhibit the photocatalytic reaction, most likely because it is shielding off the "active" components from the pollutants in the air. Consequently, it is probable that the curing must disappear from the surface, *i.e.*, under influence of traffic or weathering, before the TiO_2 will reach its optimal air purifying performance. In case of a photocatalytic spray coating, this also means that it is best to apply the TiO_2 dispersion some months after the curing compound to have the most durable effect. Alternatively, the exposed aggregates concrete can be covered with a plastic sheet to prevent dehydration (in case the concrete surface is denuded).

Figure 10. (**a**) Application of curing compound on fresh concrete, and (**b**) Effect of curing compound on photocatalytic efficiency (different samples A–D, with photocatalytic TiO_2 in mass and/or applied as dispersion, with and without curing compound).

(**a**)

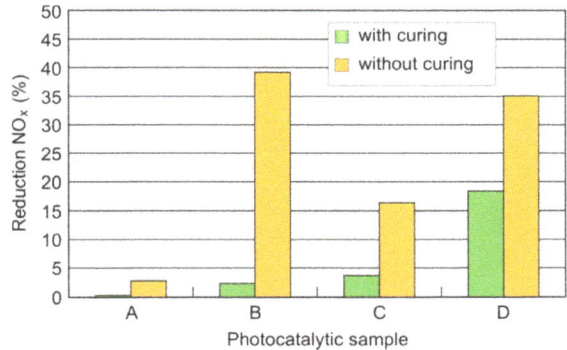

(**b**)

More detailed results of the laboratory research can be found in [28] and [29]. Based on the findings and the optimization of the concrete composition, a proper selection of photocatalytic materials and application techniques could be made, for the construction of double layered, photocatalytic concrete roads on the industrial zone "Den Hoek 3" in Wijnegem.

Double Layered Concrete at "Den Hoek 3" in Wijnegem

The concrete pavement of the industrial zone in Wijnegem has been constructed between the 15th and 18th of March 2011. The concrete was placed in two layers, wet-in-wet, with an interval time of approximately 1 hour. The bottom layer had a thickness of 180 mm, while the top layer was designed to be 50 mm. For the concrete of the bottom layer, 57% of the coarse aggregates were replaced by recycled concrete aggregates. For the top layer with TiO_2, commercially available white cement with 4% TiO_2 pre-mixed (by weight) was applied (CBR, Belgium, Heidelberg Cement Group). Two slip form pavers were used to place the concrete. As can be seen in Figure 11a, the color of the top layer is much lighter, due to the use of white cement and the presence of the TiO_2 (about 0.8 wt% of the top layer).

More information on the concrete composition, the execution and the results obtained in the lab as well as on site can be found in [28] and [29]. Besides the photocatalytic concrete roads, photocatalytic pavement blocks were also used for the bicycle lanes, parking spaces and foot paths.

Since this was a completely new industrial zone, it was not possible to have measurements on site before putting the photocatalytic concrete in place. An overview of the project is given in Figure 12. Immediately after concreting, a retarding agent was sprayed on the surface to be able to wash out the top surface after 24 h, to create an exposed aggregates surface finish (see Figure 11b). In order to prevent dehydration of the concrete during the first days, some parts of the road have been treated with curing compound; the other zones were covered with a plastic sheet. This way, the influence of the curing compound on the short and long term photocatalytic efficiency could be investigated.

Figure 11. (**a**) Construction of double layered concrete pavement at industrial zone "Den Hoek 3" in Wijnegem; (**b**) Detail of exposed aggregates surface finish of the top layer.

(**a**)   (**b**)

Figure 12. (**a**) Situation plan of the new industrial zone "Den Hoek 3" in Wijnegem, Belgium (Google Maps); (**b**) "On site" testing of photocatalytic efficiency.

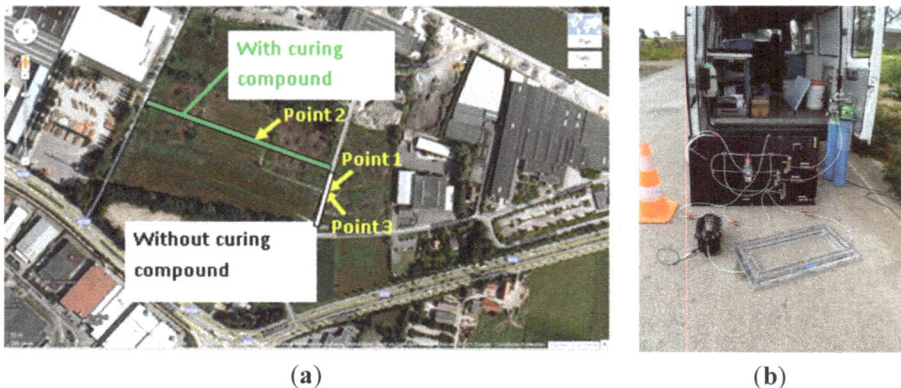

(**a**)   (**b**)

The photocatalytic efficiency of the top layer was measured in two ways: in the laboratory on cores taken from the surface at the places indicated in Figure 12a, and "on site" with a special measuring set-up, shown in Figure 12b. This "on site" test is developed to evaluate the photocatalytic properties of the concrete pavement over time and to compare the different sites (with and without curing, for example). It does not measure the overall purification of the air around the pavement but enables to measure the durability of the photocatalytic efficiency.

The set-up consists of a Plexiglas frame, screwed air-tight on the test surface (concrete pavement), and is covered with a UV-transparent glass lid. The input NO-concentration (1 ppmv), relative humidity (50% RH) and air flow (3 L/min) are taken similar to the laboratory set-up. However, the total area covered by the box is somewhat larger (700 × 300 mm²) to have a representative surface, and natural (varying!) sunlight is used in first instance to activate the surface. First results of the measurements on site are given in Figures 13 and 14, and were collected 5 months after the placement of the concrete (August 2011) at the places indicated in Figure 12a (points 1 and 2).

Figure 13. NO_x concentration measured at the outlet for zone with curing compound, 5 months after concreting (point 2, August 2011).

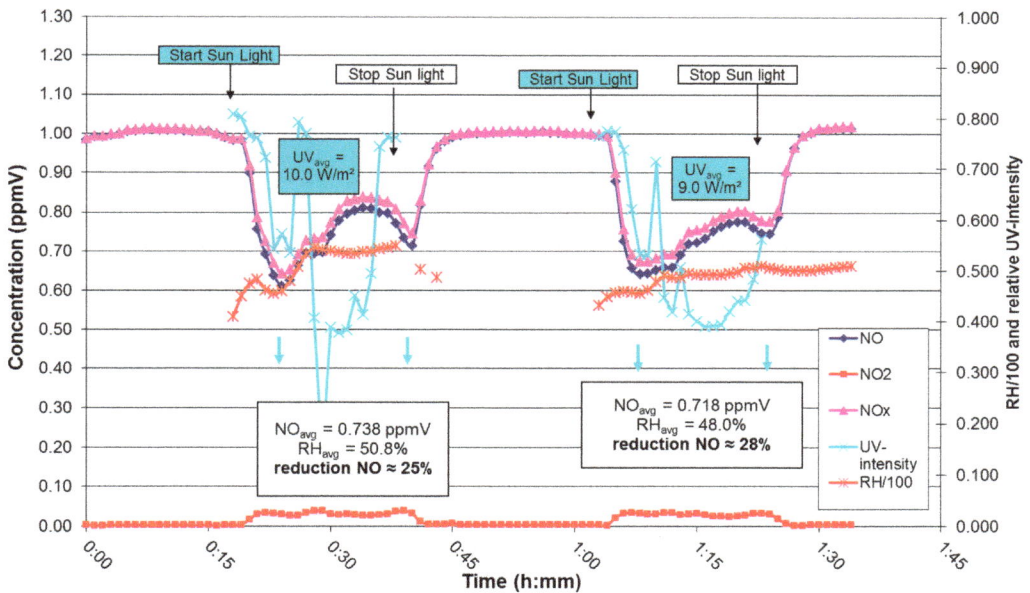

Figure 14. NO$_x$ concentration measured at the outlet for zone without curing compound, 5 months after concreting (point 1, August 2011).

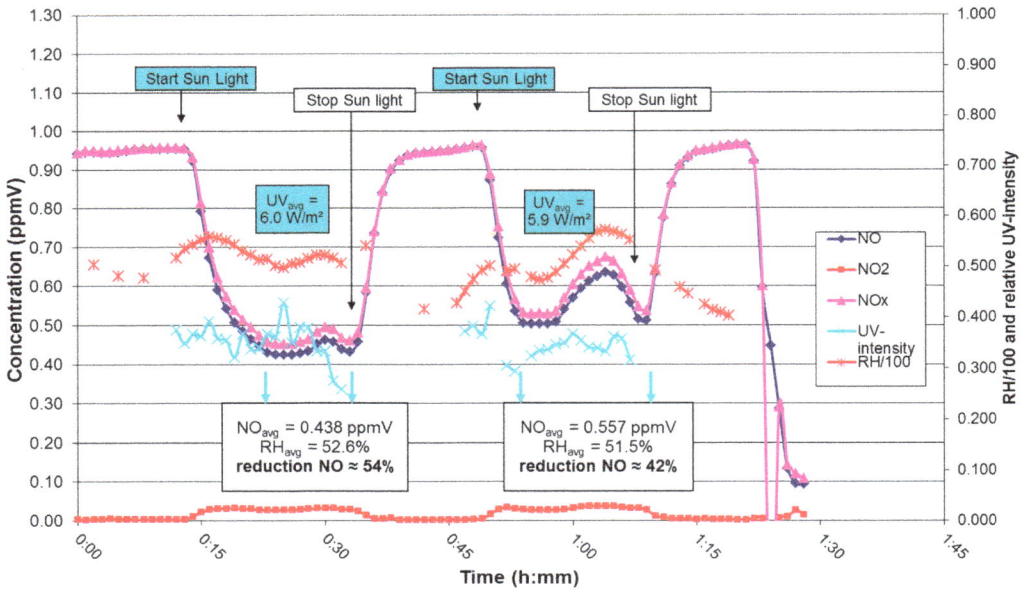

First of all, the results shown in Figures 13 and 14 indicate a large influence of the relative humidity (red curves). The NO$_x$ abatement is lower when the relative humidity increases and higher when RH decreases again. The influence of the sun light intensity (measured through the UV intensity, light blue lines) is also visible, but on a different scale: variations over a shorter period of time do not influence the NO$_x$ concentration immediately; it is the average sun light intensity over a longer period that is determining the attained NO$_x$ abatement for the photocatalytic process.

Furthermore, the reduction in NO concentration is significantly lower for the zone with curing compound, indicating it is still (slightly) inhibiting the reaction: average reduction of 27% (with curing) *versus* 48% (without curing). Nevertheless, the effect of applying a curing compound on the fresh concrete (to protect against dehydration) seems to diminish over time. These results obtained on site (year 2011) are also in line with the results obtained in the laboratory, taking into account the difference in surface, relative humidity and light intensity [28].

In order to correctly compare the results between the lab and the field, the photocatalytic activity for NO$_x$ (= sum of NO and NO$_2$) is expressed in terms of the photocatalytic deposition velocity in [m/h] under the assumption of a first order uptake kinetics and negligible transport limitations from the gas phase to the solid surface [30]:

$$k_R = \ln\left({c_0}\Big/{c_t} \right) {F}\Big/{A} \qquad (4)$$

where c_0 and c_t are the reactant concentration at the inlet and exit of the photo-reactor, respectively. In fact, this parameter refers to a first order reaction rate coefficient independent of the applied flow rate F and the active surface (A) to volume ratio of the used reactor (lab or on site). In the lab work [28], average values for the NO_x deposition velocity $k_{R,NOx}$ of 0.25 and 0.70 m/h were obtained with and without curing compound respectively, which is in nice agreement with the results on site for 2011 (see further in Table 1).

The measurements on site are also repeated over time in order to see the influence of ageing and traffic on the photocatalytic efficiency. Recent measurements performed in the summer of 2012 for instance, are shown in Figure 15. Here, measurements were performed using an external UV-lamp (10 W/m²) as well as natural sunlight to activate the photocatalyst present in the pavement. It appears the activity under sun light is somewhat higher compared to the UV lamp only. This could be due to the fact that the applied TiO_2 (in the active cement) is also partially active under visible light and/or is excited by the shorter UV wavelengths (UV-B, UV-C) present in the sun spectrum.

Figure 15. NO_x concentration measured at the outlet for zone without curing compound, 17 months after concreting (point 3, August 2012).

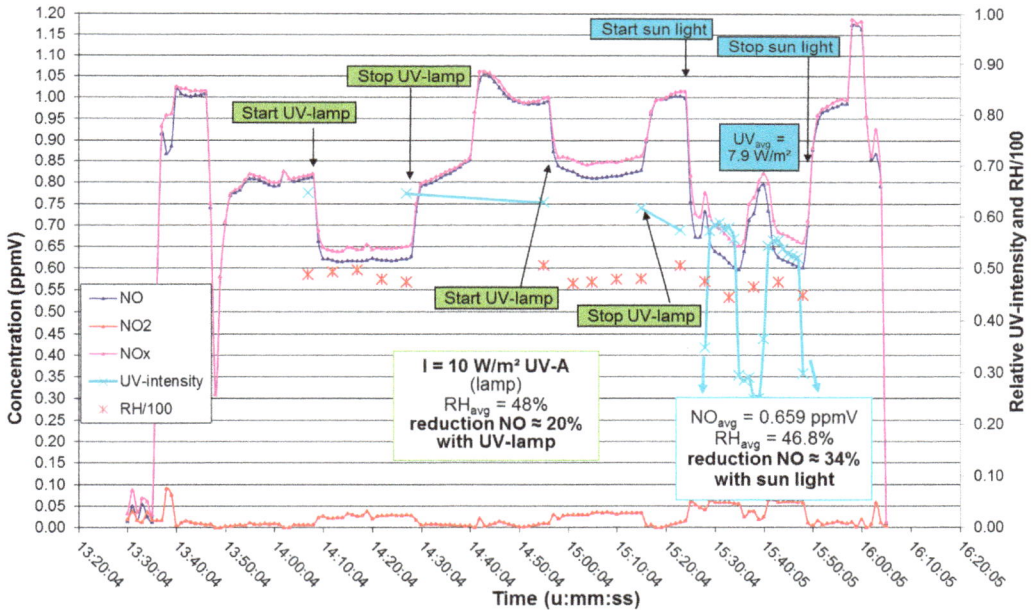

On the other hand, the measurement of the UV-intensity comprised in the sun light could be erroneous because of the radiometer used here. This is only calibrated for specific UV-A lamps (between 300 and 400 nm) applied in the geometry of the lab set-up which differs substantially from these exterior tests. The activity observed under natural, varying sun light though, is still very interesting from the view point of practical application. The use of the external UV-lamp with constant light intensity in turn, allows making a more absolute comparison of the photocatalytic activity between different zones and for different times.

In any case, the results of Figure 15 reveal already that the efficiency of this kind of photocatalytic application (TiO_2 integrated in the cement) appears to decrease in time: on average 34% NO-reduction (after 17 months) *versus* 48% (after 5 months). Possible causes could reside in the covering of the TiO$_2$ at the surface by dirt, the detachment of the TiO$_2$ from the surface or the deposition of products from chemical reactions which can take place at the surface.

In this respect, in October 2012 an aqueous TiO$_2$ dispersion (Eoxolit® consisting of a mixture of two different types of TiO$_2$ particles with a total concentration of 40 g/L TiO$_2$) was also applied on the surface in some parts of the roads on the industrial zone in Wijnegem, as shown in Figure 16a, for the purpose of comparative measurements. In total four different zones were considered:

- Zone 1 = double layered concrete (0/6.3 mm on top) without TiO$_2$;
- Zone 2 = single layered concrete (0/20 mm) without TiO$_2$;
- Zone 3 = double layered concrete with TiO$_2$ (active cement) and without curing compound;
- Zone 4 = double layered concrete with TiO$_2$ (active cement) and with curing compound.

The photocatalytic dispersion was applied with a dose of approximately 1 L per 5 m^2 on a total of 800 m^2, followed by spraying of a hydrophobic product for optimal functioning of the coating (manufacturers' guidelines). Important to mention however, is the fact that at the time of application there was a severe pollution with soil and dirt at the surface of the pavement in some zones due to the presence of a grinding installation plant at the site. This most certainly had an impact on the efficiency of the TiO$_2$ suspension (see further). Subsequently, provisional controls of the photocatalytic efficiency have been carried out to check the separate action of the two types of photoactive materials (mass and dispersion), and to further assess the durability of the air purifying performance. Most recent measurements on the site in Wijnegem were performed in the summer of 2013, at the measurement points (1–9) indicated in Figure 16b. All results obtained up till now (2011–2013) are summarized in Table 1.

Figure 16. (**a**) Application of photocatalytic dispersion on part of the roads at industrial zone "Den Hoek 3" (October 2012); (**b**) Localization of measurement points for "on site" testing (Google Maps).

(**a**)   (**b**)

Table 1. Summary of results in time for photocatalytic activity measured on site in Wijnegem.

Zone	$k_{R,NO}$ ($k_{R,NOx}$) [m/h]				
	Sun light			UV-lamp (10 W/m²)	
	2011	2012	2013	2012	2013
4: **with** curing compound, active cement (point 2, 4)	0.30 (0.26)	0.09 (0.07)	–	0.06 (0.04)	–
3: **without** curing compound, active cement (point 1 and 3)	0.70 (0.66)	0.39 (0.34)	0.38 (0.28)	0.21 (0.19)	0.22 (0.18)
4: **with** curing, active cement +TiO$_2$ dispersion (point 4)	–	–	0.82 (0.62)	–	0.28 (0.22)
3: **without** curing, active cement +TiO$_2$ dispersion (points 6 and 9)	–	–	0.27 (0.20)	–	0.21 (0.15)
1: double layered, **no** active cement + TiO$_2$ dispersion (point 7)	–	–	0.32 (0.27)	–	0.15 (0.13)
2: single layered, **no** active cement + TiO$_2$ dispersion (point 8)	–	–	0.14 (0.13)	–	0.08 (0.07)

First of all, when comparing the measurements on the surface of the pavement at points 1 and 3 (*cf.* Figure 16b) in 2013 with these of 2012, a very similar result can be noticed: a photocatalytic deposition velocity for NO$_x$ of *ca.* 0.2 m/h under UV light. This indicates that the decreasing trend in photocatalytic activity for the concrete with "active" cement (see evolution 2011–2012) seems to be stabilized in 2013.

Furthermore, the measured efficiency for points 1 and 3 (in 2013) appears to differ little or nothing with the one for points 6 and 9, with application of the photocatalytic coating (TiO$_2$ dispersion) on the pavement surface. Here, the TiO$_2$ dispersion did not produce a significant added value (yet) in terms of photocatalytic air purifying action. Only for point 4 (active cement with curing compound, after application of TiO$_2$ dispersion) one can notice a strong improvement of the photocatalytic efficiency (deposition velocity of *ca.* 0.8 m/h for NO under sun light and nearly 0.3 m/h under UV). Possibly, the pollution of the surface at the time of application has played an important part causing the adhesion of the coating to be far from optimal.

For points 7 and 8 (concrete without active cement, but with TiO$_2$ dispersion on the surface), the activity is not significantly better either (or even less) compared to the "pure" concrete with active cement. In addition, point 8 (single layered concrete 0/20) reveals much smaller photocatalytic reactivity than point 7 (double layered concrete with top layer 0/6.3): deposition velocity of 0.08 m/h *versus* 0.15 m/h for NO reduction under UV. This probably has to do with the stronger adhesion of the coating on the surface of the finer (0/6.3) double layered concrete compared to the coarser (0/20) single layered concrete.

Finally, a measurement on site was also performed for the newly constructed pavements at the industrial zone in Lier, which have a different surface finishing as illustrated in Figure 17a. The results of this measurement, 20 months after construction, are shown in Figure 17b.

194

Figure 17. (a) Double layered photocatalytic concrete pavement with brushed surface finish at industrial zone "Duwijckpark" in Lier; **(b)** NO_x concentration measured at the outlet for the site in Lier (active cement + curing compound), 20 months after concreting (August 2013).

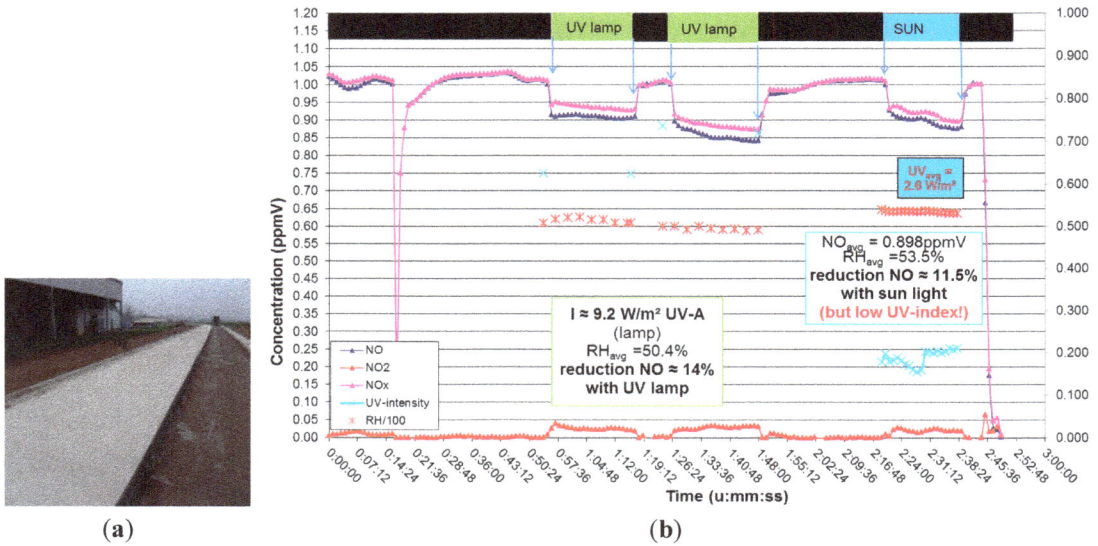

(a)   (b)

In comparison with the measurements of Wijnegem in 2013 (see Table 1), a slightly lower photocatalytic reaction is observed in Lier, which among others is due to the use of a curing compound (for the brushed surface) and the lower TiO_2 content (less cement used). However, if we make the comparison with the zone with curing compound in Wijnegem (*cf.* point 4 in zone 4) measured in 2012 (17 months after construction), a significantly better result under UV light is obtained in Lier: deposition velocity for NO of 0.14 m/h in Lier *versus* 0.06 m/h in Wijnegem. This higher activity probably has to do with the brushed surface finish instead of exposing the aggregates (*cf.* Figure 9). In any case, these measurements confirm the photocatalytic action 20 months after construction of the concrete pavement.

6. Conclusions and Perspectives

Photocatalytic (TiO_2 containing) paving materials with the potential of reducing air pollution by traffic are being used more frequently on site in horizontal as well as in vertical applications, also in Belgium. Laboratory results indicate a good efficiency towards the abatement of NO_x in the air by using these innovative materials. The durability of the photocatalytic action also remains mostly intact, though regular cleaning (by rain) of the surface is necessary. The relative humidity (RH) is an important parameter, which may reduce the efficiency on site. If the RH is too high, the water will be adsorbed at the surface and prevent the reaction with the pollutants.

The translation from the laboratory results to the "on-site efficiency" is still a difficult and critical factor, because of the great number of parameters involved. Hence, there is still a need for large scale applications to demonstrate the effectiveness of photocatalytic materials in "real life" and evaluate the durability of the air purifying action, such as the European Life+ project PhotoPAQ and the industrial zones "Den Hoek 3" in Wijnegem en "Duwijckpark" in Lier. These recent applications in Belgium show already some interesting results.

It seems the use of photocatalytic cement-based coatings inside road tunnels is not mature for application on a large scale yet. From the experience gained during the Leopold II tunnel campaigns in Brussels, recommendations for the proper use of these innovative materials can be made though, such as:

- Optimized coating application for low surface roughness and minimizing dust adsorption;
- High UV light intensity levels in the order of magnitude of 10 W/m²;
- Low average relative humidity of tunnel air ($\leq 60\%$);
- High enough photocatalytic activity, with threshold values defined from lab studies;
- Low average wind speed (< 2 m/s) in the tunnel for increased reaction time of pollutants;
- High surface to volume ratio (smaller sized tunnel tubes).

For the double layered photocatalytic concrete pavements using active cement, an efficiency comparable to the one measured in the laboratory is obtained initially; though it seems to decrease somewhat in time due to dirt build-up and other deposits on the surface, the air purifying action has stabilized after more than two years (2011–2013). Application of a curing compound—to protect the fresh concrete against desiccation—initially strongly reduces the photocatalytic activity and also has an impact on the long term. Use of a plastic sheet to protect the young concrete is therefore recommended. Furthermore, the exposed aggregates technique is not ideal for the photocatalytic efficiency since in this case a lot of aggregates are present at the surface and the TiO_2 is only present in the paste. The application of a brushed surface finish could lead to a better result.

Use of a photocatalytic coating (TiO_2 dispersion) on the surface of the concrete pavement does not produce an added value for the air purifying action compared to mixing in the mass, despite the good results in the laboratory. This probably has to do with the loss of adhesion in time and the filthiness of the surface at the time of application. Possibly, the coating is partially washed away with the dirt. In addition, better results are obtained on the finer, double layered concrete (0/6.3) than for the coarser, single layered concrete (0/20) which could be due to the better adhesion of the coating on the surface.

Durability of the photocatalytic action in time (for products mixed in the mass and/or applied on the surface) and optimization of the adhesion of photo-active coatings on the concrete surface, are topics that need to be investigated further.

Finally, the best results will be achieved by modeling the environment, validating the models by measurements on site, followed by an implementation of the different influencing parameters to assess the real life effect. One must bear in mind that photocatalytic applications are not effective everywhere; "good" contact between the airborne pollutants and the active surface is crucial and factors such as wind speed and direction, street configuration and pollution sources all play a very important role.

196

Acknowledgements

The authors wish to thank IWT Flanders (Institute for the Promotion of Innovation by Science and Technology in Flanders), FPS Economy (Federal Public Service), Life+ and EFRO (European Union), INTERREG and the Ministry of the Brussels-Capital Region—Brussels Mobility for the (financial) support of the different projects.

Author Contributions

E.L.B and A.B. both coordinated and supervised the different research projects involving photocatalytic applications; E.L.B. prepared the manuscript. All authors read and approved the manuscript.

Conflicts of Interest

The authors declare no conflict of interest.

References

1. European Commission. *EU Energy and Transport in Figures, Statistical Pocketbook*; Publications Office of the European Union: Brussels, Belgium, 2011.
2. Beeldens, A. Air purification by pavement blocks: Final results of the research at the BRRC. In Proceedings of Transport Research Arena—TRA 2008, Ljubljana, Slovenia, 21–24 April 2008.
3. Directive 2008/50/EC of the European Parliament and of the Council on ambient air quality and cleaner air for Europe. *Off. J. Eur. Union* **2008**, L152:1–L152:44.
4. Chen, J.; Poon, C. Photocatalytic construction and building materials: From fundamentals to applications. *Build. Environ.* **2009**, *44*, 1899–1906.
5. Renz, C. Lichtreaktionen der Oxyde des Titans, Cers und der Erdsäuren. *Helv. Chim. Acta* **1921**, *4*, 961–968.
6. Fujishima, A.; Honda K. Electrochemical photolysis of water at a semiconductor electrode. *Nature* **1972**, *238*, 37–38.
7. Fujishima, A.; Rao, T.N.; Tryk, D.A. Titanium dioxide photocatalysis. *J. Photochem. Photobiol. C* **2000**, *1*, 1–21.
8. Sopyan, I.; Watanabe, M.; Murasawa, S.; Hashimoto, K.; Fujishima, A. An efficient TiO_2 thin-film photocatalyst: Photocatalytic properties in gas-phase acetaldehyde degradation. *J. Photochem. Photobiol. A* **1996**, *98*, 79–86.
9. Cassar, L.; Pepe, C. Paving Tile Comprising an Hydraulic Binder and Photocatalyst Particles. EP-Patent 1600430 A1, 1997.
10. Murata, Y.; Tawara, H.; Obata, H.; Murata, K. NO_x-Cleaning Paving Block. EP-Patent 0786283 A1, 1996.
11. Ohama, Y.; Van Gemert, D. *Application of Titanium Dioxide Photocatalysis to Construction Materials*; Springer: Dordrecht, The Netherlands, 2011.

12. Saubere Luft Durch Pflastersteine Clean Air by Airclean®. Available online: http://www.ime.fraunhofer.de/content/dam/ime/de/documents/AOe/2009_2010_Saubere%20Luft%20durch%20Pflastersteine_s.pdf (accessed on 25 July 2014).

13. Dillert, R.; Stötzner, J.; Engel, A.; Bahnemann, D.W. Influence of inlet concentration and light intensity on the photocatalytic oxidation of nitrogen(II) oxide at the surface of Aeroxide® TiO$_2$ P25. *J. Hazard. Mater.* **2012**, *211–212*, 240–246.

14. Laufs, S.; Burgeth, G.; Duttlinger, W.; Kurtenbach, R.; Maban, M.; Thomas, C.; Wiesen, P.; Kleffmann, J. Conversion of nitrogen oxides on commercial photocatalytic dispersion paints. *Atmos. Environ.* **2010**, *44*, 2341–2349.

15. Devahasdin, S.; Fan, C.; Li, J.K.; Chen, D.H. TiO$_2$ photocatalytic oxidation of nitric oxide: Transient behavior and reaction kinetics. *J. Photochem. Photobiol. A* **2003**, *156*, 161–170.

16. Ballari, M.M.; Yu, Q.L.; Brouwers, H.J.H. Experimental study of the NO and NO$_2$ degradation by photocatalytically active concrete. *Catal. Today* **2011**, *161*, 175–180.

17. Fujishima, A.; Zhang, X. Titanium dioxide photocatalysis: Present situation and future approaches. *Comptes Rendus Chim.* **2006**, *9*, 750–760.

18. PhotoPAQ (2010–2014) Life+ Project. Available online: http://photopaq.ircelyon.univ-lyon1.fr/ (accessed on 25 July 2014).

19. *ISO 22197-1:2007 Fine Ceramics (Advanced Ceramics, Advanced Technical Ceramics)—Test Method for Air-Purification Performance of Semi Conducting Photocatalytic Materials—Part 1: Removal of Nitric Oxide*; International Standards Organization (ISO): Geneva, Switzerland 2007.

20. CEN Technical Committee 386 "Photocatalysis" Business Plan—(internet) Draft BUSINESS PLAN CEN/TC386 PHOTOCATALYSIS. Available online: http://standards.cen.eu/BP/653744.pdf (accessed on 28 July 2014).

21. Hüsken, G.; Hunger, M.; Brouwers, H.J.H. Experimental study of photocatalytic concrete products for air purification. *Build. Environ.* **2009**, *44*, 2463–2474.

22. Beeldens, A.; Boonen, E. Photocatalytic applications in Belgium, purifying the air through the pavement. In Proceedings of the XXIVth World Road Conference, Mexico City, Mexico, 26–30 September 2011.

23. Maggos, Th.; Plassais, A.; Bartzis, J.G.; Vasilakos, Ch.; Moussiopoulos, N.; Bonafous, L. Photocatalytic degradation of NO$_x$ in a pilot street canyon configuration using TiO$_2$-mortar panels. *Environ. Monit. Assess.* **2008**, *136*, 35–44.

24. Gignoux, L.; Christory, J.P.; Petit, J.F. Concrete roadways and air quality—Assessment of trials in Vanves in the heart of the Paris region. In Proceedings of the 12th International Symposium on Concrete Roads, Sevilla, Spain, 13–15 October 2010.

25. Guerrini, G.L. Photocatalytic performances in a city tunnel in Rome: NO$_x$ monitoring results. *Constr. Build. Mater.* **2012**, *27*, 165–175.

26. Boonen, E.; Akylas, V.; Barmpas, F.; Boréave, A.; Bottalico, L.; Cazaunau, M.; Chen, H.; Daële, V.; De Marco, T.; Doussin, J.F.; *et al.* Photocatalytic de-pollution in the Leopold II tunnel in Brussels, Part I: Construction of the field site. *Constr. Build. Mater.* **2014**, Submitted.

27. Gallus, M.; Akylas, V.; Barmpas, F.; Beeldens, A.; Boonen, E.; Boréave, A.; Bottalico, L.; Cazaunau, M.; Chen, H.; Daële, V.; *et al.* Photocatalytic de-pollution in the Leopold II tunnel in Brussels, Part II: NO$_x$ abatement results. *Constr. Build. Mater.* **2014**, Submitted.
28. Boonen, E.; Beeldens, A. Photocatalytic roads: From lab testing to real scale applications. *Eur. Transp. Res. Rev.* **2013**, *5*, 79–89.
29. Beeldens, A.; Boonen, E. A double layered photocatalytic concrete pavement: A durable application with air-purifying properties. In Proceedings of 10th International Conference on Concrete Pavements (ICCP), Quebec, Canada, 8–12 July 2012.
30. Ifang, S.; Gallus, M.; Liedtke, S.; Kurtenbach, R.; Wiesen, P.; Kleffmann, J. Standardization methods for testing photo-catalytic air remediation materials: Problems and solution. *Atmos. Environ.* **2014**, *91*, 154–161.

MDPI AG
Klybeckstrasse 64
4057 Basel, Switzerland
Tel. +41 61 683 77 34
Fax +41 61 302 89 18
http://www.mdpi.com/

Coatings Editorial Office
E-mail: coatings@mdpi.com
http://www.mdpi.com/journal/coatings

www.ingramcontent.com/pod-product-compliance
Lightning Source LLC
Chambersburg PA
CBHW051921190326
41458CB00026B/6360